An OPUS book

THE PROBLEMS OF MATHEMATICS

Ian Stewart is Professor of Mathematics at Warwick University. He is the author of *Does God Play Dice?*, *Game, Set and Math*, and *Fearful Symmetry: Is God a Geometer?* He writes the 'Mathematical Recreations' column of *Scientific American*.

OPUS General Editors

Walter Bodmer
Christopher Butler
Robert Evans
John Skorupski

OPUS books provide concise, original, and authoritative introductions to a wide range of subjects in the humanities and sciences. They are written by experts for the general reader as well as for students.

The Problems of Mathematics

IAN STEWART

Second Edition

Oxford   New York

OXFORD UNIVERSITY PRESS

1992

Oxford University Press, Walton Street, Oxford OX2 6DP
Oxford New York Toronto
Delhi Bombay Calcutta Madras Karachi
Petaling Jaya Singapore Hong Kong Tokyo
Nairobi Dar es Salaam Cape Town
Melbourne Auckland
and associated companies in
Berlin Ibadan

Oxford is a trade mark of Oxford University Press

British Library Cataloguing in Publication Data
Data available
ISBN 0-19-219262-0
ISBN 0-19-286148-4 Pbk

Library of Congress Cataloging in Publication Data
Stewart, Ian.
The problems of mathematics/Ian Stewart.—2nd ed.
p. cm.
"An OPUS book"—Half t.p.
Includes bibliographical references and index.
1. Mathematics—Popular works. I. Title.
510—dc20 QA93.S74 1992 92-6143
ISBN 0-19-219262-0
ISBN 0-19-286148-4 Pbk

Typeset by Best-set Typesetter Ltd., Hong Kong

Foreword

Mathematics is not a book confined within a cover and bound between brazen clasps, whose contents it needs only patience to ransack; it is not a mine, whose treasures may take long to reduce into possession, but which fill only a limited number of veins and lodes; it is not a soil, whose fertility can be exhausted by the yield of successive harvests; it is not a continent or an ocean, whose area can be mapped out and its contour defined: it is as limitless as that space which it finds too narrow for its aspirations; its possibilities are as infinite as the worlds which are forever crowding in and multiplying upon the astronomer's gaze; it is as incapable of being restricted within assigned boundaries or being reduced to definitions of permanent validity, as the consciousness of life, which seems to slumber in each monad, in every atom of matter, in each leaf and bud cell, and is forever ready to burst forth into new forms of vegetable and animal existence.

JAMES JOSEPH SYLVESTER

Foreword to the second edition

It's *déjà vu* all over again.

YOGI BERRA

Contents

List of Figures

Interview with a mathematician . . .

conducted by Seamus Android
on behalf of the
proverbial man-in-the-street

Tell the truth, but tell it slant.

Emily Dickinson

Android:	Good evening, viewers, and welcome yet again to *Boffins and Brains*, the show where scientists let their hair down and tell us what they're up to. Tonight we have a mathematician in the studio. [Turns to mathematician.] Welcome to the show.
Mathematician:	Thank you, Seamus. Hello everybody.
Android:	In previous programmes we've had a physicist telling us how atoms bounce off a crystal; a chemist talking about new kinds of plastic; a biologist on the early development of the giraffe embryo; an engineer with some new ideas on public transport; and an astronomer who's just discovered the most distant galaxy in the known universe. So, Mathematician: what delights do you have in store for us?
Mathematician:	I thought I'd say a bit about how you can get a TOP but non-DIFF 4-manifold by surgery on the Kummer surface. You see, there's this fascinating cohomology intersection form related to the exceptional Lie algebra E_8, and . . .
Android:	[Sarcastically] That's fascinating.
Mathematician:	[Surprised and pleased] Thank you.
Android:	Is all that gobbledegook really significant?

Mathematician:	[Surprised but no longer pleased] Of course! It's one of the most important discoveries of the last decade!
Android:	Can you *explain* it in words ordinary mortals can understand?
Mathematician:	Look, buster, if ordinary mortals could understand it, you wouldn't need mathematicians to do the job for you, right?
Android:	I don't want the technical details. Just a general feeling for what's going on.
Mathematician:	You can't get a feeling for what's going on without understanding the technical details.
Android:	Why not?
Mathematician:	[Who has never thought about it before] Well, you just can't.
Android:	Physicists seem to manage.
Mathematician:	But they work with things from everyday experience.
Android:	Sure. 'How gluon antiscreening affects the colour charge of a quark.' 'Conduction bands in Gallium Arsenide.' Trip over 'em all the time on the way to work, don't you?
Mathematician:	Yes, but . . .
Android:	I'm sure that the physicists find all the technical details just as fascinating as you do. But they don't let them intrude so much.
Mathematician:	But how can I explain things properly if I don't give the details?
Android:	How can anyone else understand them if you do?
Mathematician:	But if I skip the fine points, some of the things I say won't be completely true! How can I talk about manifolds without mentioning that the theorems only work if the manifolds are finite-dimensional paracompact Hausdorff with empty boundary?
Android:	Lie a bit.
Mathematician:	Oh, but I couldn't do that!
Android:	Why not? Everybody *else* does.
Mathematician:	[Tempted, but struggling against a life-

	time's conditioning] But I *must* tell the truth!
Android:	Sure. But you might be prepared to bend it a little, if it helps people understand what you're doing.
Mathematician:	[Uncertain] Well . . .
Android:	Let's start over. What's the main thing mathematics is about?
Mathematician:	Solving problems.
Android:	Fine. So tell us what the problems are, where they come from, how they get solved, what the people who solve them are like, what you can do with the answers when you've got them, what problems haven't been solved yet, how solving them or failing to solve them changes people's view of what mathematics is and where it's going. That sort of thing. The whole of mathematics in a nutshell.
Mathematician:	Oh, I can't cover the whole thing! It's much too big. I'll do the best I can, but there are lots of areas I don't know the foggiest thing about, and I'll have to ignore those.
Android:	Fair enough. It'll be a revelation to a lot of people that anything new ever happens in mathematics at all.
Mathematician:	And I can't avoid technicalities altogether. I mean, I can't discuss research in group theory without saying roughly what a group is. Any more than a physicist could talk about black holes without saying something about curved space-time.
Android:	Fair enough. But don't overdo it. I mean . . .
Mathematician:	. . . don't explain how the assumption of a C^r atlas on a pseudo-Riemannian manifold implies the existence of an analytic atlas.
Android:	I couldn't have put it better myself.
Mathematician:	[Sceptical, but excited at his own daring] Well, I suppose I could give it a *try* . . .

Interview to the second edition

with anchorperson Virginia Plain, for the
Rough Guide to Mathematics

When you come to a fork in the road—take it.

Yogi Berra

Virginia Plain: You know, you're pushing your luck with this one. A lot of them didn't understand Seamus Android.

Mathematician: He's a character from *Round the Horne*, a sixties radio comedy.

Virginia Plain: They thought it was an Irish joke. In poor taste.

Mathematician: I suppose it was, come to think of it. But not mine. We had a *cat* called Seamus Android, you know ... Still, I should be safe being interviewed by a character from a Roxy Music record.

Virginia Plain: Now *I'm* getting confused. Look, what's this interview for?

Mathematician: You're the anchorperson, you tell me. I just got a message to turn up at the studio.

Virginia Plain: [Consults teleprompt.] You wrote this book, all about how mathematics has problems. Don't we all? Still, surprise, it sold. Now they want another one, right? Sequel, prequel, trilogy, or dekalogy?

Mathematician: Just a second edition.

Virginia Plain: What was wrong with the first one?

Mathematician: It got out of date.

Virginia Plain: How long did that take? A century?

Mathematician: Five years.

Virginia Plain: *Five years?* You mean so much new math-

	ematics happened in five years that you had to do a *new book*?
Mathematician:	Not completely new, Virginia. Three or four new chapters, odd sections here and there, polish up a few sentences, bring odd remarks up to date. Mathematics is moving fast nowadays.
Virginia Plain:	So I see. Wow. I though it was all polished off long ago. Tablets of stone. 'The answers are all in the back of the book.'
Mathematician:	That's because most of the questions aren't in the front.
Virginia Plain:	*Touché.* So, like, what's new in math?
Mathematician:	Well . . . Kepler's sphere-packing problem, for a start. Laczkovich's solution to Tarski's circle-squaring problem. The Jones polynomial, a big breakthrough in knot theory. Ed Witten's work linking it to quantum field theory.
Virginia Plain:	What are those about?
Mathematician:	The Kepler Problem dates back to 1611. You have to prove that the best way to stack spheres is the same one that green-grocers use. It's lasted 380 years, but now somebody claims to have a proof.
Virginia Plain:	So now greengrocers can sleep safely in their beds at night, knowing that they're not wasting space when they put out their oranges?
Mathematician:	Sure. And physicists have a better idea why atoms stack into crystal lattices. Don't forget that mathematics is very general and very powerful. Mind you, not everybody is convinced the proof is right.
Virginia Plain:	What's with this guy Laczkovich? I thought squaring the circle was impossible.
Mathematician:	It is, with the traditional tools of ruler and compasses. Which actually aren't quite as traditional as people tend to think, and that's something else I've tried to fix this

	time. Alfred Tarski asked a rather different question: whether you can cut a square into finitely many pieces and reassemble them to get a circle.
Virginia Plain:	Sounds unlikely. What about the curved edges?
Mathematician:	Laczkovich made them disappear. He had to use fractal pieces, of course. About 10^{50} of them.
Virginia Plain:	I won't ask to see a picture, then. I'll take your word for it. By the way, I'm not sure I approve of your 'Save the Krill' T-shirt.
Mathematician:	Uh—let me tell you about the Jones polynomial, Virginia. That came as a total surprise. It's a new and better way to tell knots apart. It started in mathematical physics, wandered all over topology and algebra, and now it's gone back into physics again. With some molecular biology on the side. It's one of the hottest properties around.
Virginia Plain:	Anything else?
Mathematician:	Quantum cryptography. I call it Schrödinger's cat-flap.
Virginia Plain:	What's that about?
Mathematician:	How to use quantum indeterminacy to transmit messages with a guarantee nobody's tapping the line.
Virginia Plain:	Cat-flap?
Mathematician:	Read the book if you want to know.
Virginia Plain:	This is the BBC, we don't allow commercials.
Mathematician:	I bet *that* bit's going to look dated in five years' time . . . sorry, got carried away. I won't do it again, I promise. What else is new? Lots, so much that I had to leave a lot of nice stuff out. But I've managed to include discrete graphics, symmetric chaos, infinitely fast computers, and the uncomputability of the Mandelbrot set.

Virginia Plain:	Wow, I really dig that crazy Mandelbrot set. I've got a Mandelbrot track-suit, you know—I wear it to my vegan aerobics classes. What do you mean, uncomputable?
Mathematician:	There's no way to decide whether a point is in it or not.
Virginia Plain:	Come off it. How do they draw those wonderful pictures, then?
Mathematician:	They're just approximations. You can get a really good approximation if you calculate for long enough. But you always get the fine detail wrong.
Virginia Plain:	Anything else new?
Mathematician:	Yes, Elkies' counter-example to Euler's conjecture. Three perfect fourth powers that add up to a perfect fourth power. Lots of odds and ends here and there, like the abc conjecture in number theory... Woops, nearly forgot: digital sundials!
Virginia Plain:	Aren't sundials a bit old-fashioned?
Mathematician:	Yes, but these are *digital*. State-of-the-art. Hang 'em up in the sunshine and read the time off the wall—in digits. It's a theorem about the restrictions obeyed by projections of fractal sets, really.
Virginia Plain:	Really. And what restrictions do they obey?
Mathematician:	None whatsoever.
Virginia Plain:	Are you winding me up? I sure hope there's a sanity clause in your publishing contract.
Mathematician:	Sorry, Virginia, there is no sanity clause.

1
The nature of mathematics

Go down I times three, one third of me, one
fifth of me is added to me; return I, filled am I.
What is the quantity saying it?

A'h-mosé The Scribe—Rhind Papyrus

One of the biggest problems of mathematics is to explain to
everyone else what it is all about. The technical trappings
of the subject, its symbolism and formality, its baffling terminology, its apparent delight in lengthy calculations: these tend
to obscure its real nature. A musician would be horrified if his
art were to be summed up as 'a lot of tadpoles drawn on a row
of lines'; but that's all that the untrained eye can see in a
page of sheet music. The grandeur, the agony, the flights of
lyricism and the discords of despair: to discern them among
the tadpoles is no mean task. They are present, but only in
coded form. In the same way, the symbolism of mathematics
is merely its coded form, not its substance. It too has its
grandeur, agony, and flights of lyricism. However, there is a
difference. Even a casual listener can enjoy a piece of music.
It is only the performers who are required to understand the
antics of the tadpoles. Music has an immediate appeal to
almost everybody. But the nearest thing I can think of to a
mathematical performance is the Renaissance tournament,
where leading mathematicians did public battle on each other's
problems. The idea might profitably be revived; but its appeal
is more that of wrestling than of music.

Music can be appreciated from several points of view: the
listener, the performer, the composer. In mathematics there is
nothing analogous to the listener; and even if there were, it
would be the composer, rather than the performer, that would
interest him. It is the creation of new mathematics, rather

than its mundane practice, that is interesting. Mathematics is not about symbols and calculations. These are just tools of the trade—quavers and crotchets and five-finger exercises. Mathematics is about *ideas*. In particular it is about the way that different ideas relate to each other. If certain information is known, what else must necessarily follow? The aim of mathematics is to understand such questions by stripping away the inessentials and penetrating to the core of the problem. It is not just a question of getting the right answer; more a matter of understanding why an answer is possible at all, and why it takes the form that it does. Good mathematics has an air of economy and an element of surprise. But, above all, it has *significance*.

Raw materials

I suppose the stereotype of the mathematician is an earnest, bespectacled fellow poring over an endless arithmetical screed. A kind of super-accountant. It is surely this image that inspired a distinguished computer scientist to remark in a major lecture that 'what can be done in mathematics by pencil and paper alone has been done by now'. He was dead wrong, and so is the image. Among my colleagues I count a hang-gliding enthusiast, a top-rank mountaineer, a smallholder who can take a tractor to bits before breakfast, a poet, and a writer of detective stories. And none of them is especially good at arithmetic.

As I said, mathematics is not about calculations but about ideas. Someone once stated a theorem about prime numbers, claiming that it could never be proved because there was no good notation for primes. Carl Friedrich Gauss proved it from a standing start in five minutes, saying (somewhat sourly) 'what he needs is *notions*, not *notations*'. Calculations are merely a means to an end. If a theorem is proved by an enormous calculation, that result is not properly understood until the reasons why the calculation works can be isolated and seen to appear as natural and inevitable. Not all ideas are mathematics; but all good mathematics must contain an idea.

Pythagoras and his school classified mathematics into four branches, like this:

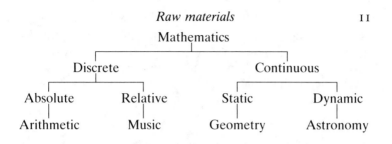

Three of these branches remain major sources of mathematical inspiration. The fourth, music, is no longer given the same kind of prominence, but it can be re-interpreted as the algebraic or combinatorial approach. (There is still a pronounced tendency for mathematicians to be musically talented.) To these four, modern mathematics has added a fifth: Lady Luck. Thus there are now at least five distinct sources of mathematical ideas. They are number, shape, arrangement, movement, and chance.

The most basic and well known is number. Originally the number concept must have arisen through counting: possessions, days, enemies. Measurement of lengths and weights led to fractions and the 'real' numbers. A major act of mathematical imagination created 'imaginary' numbers such as $\sqrt{-1}$. From that point on mathematics was never quite the same. Shape, or form, leads to geometry: not just the stereotyped and pedantic style of geometry that used to be blamed on Euclid, but its modern offspring such as topology, singularity theory, Lie groups, and gauge field theories. Novel geometric forms—fractals, catastrophes, fibre bundles, strange atrractors—still inspire new developments in mathematics. Problems about ways to arrange objects according to various rules lead to combinatorics, parts of modern algebra and number theory, and what is becoming known as 'finite mathematics', the basis of much computer science. Movement—of cannonballs, planets, or waves—inspired the calculus, the theories of ordinary and partial differential equations, the calculus of variations, and topological dynamics. Many of the biggest areas of mathematical research concern the way systems evolve in time. A more recent ingredient is chance, or randomness. Only for a couple of centuries has it been realized that chance has its own type of pattern and regularity;

only in the last fifty years has it been possible to make this statement precise. Probability and statistics are obvious results; less well known but equally important is the theory of stochastic differential equations—dynamics plus random interference.

The driving force

The driving force of mathematics is *problems*. A good problem is one whose solution, rather than merely tidying up a dead end, opens up entirely new vistas. Most good problems are hard: in mathematics, as in all walks of life, one seldom gets something for nothing. But not all hard problems are good: intellectual weight-lifting may build mental muscles, but who wants a muscle-bound brain? Another important source of mathematical inspiration is *examples*. A really nice self-contained piece of mathematics, concentrating on one judiciously chosen example, often contains within it the germ of a general theory, in which the example becomes a mere detail, to be embellished at will. I'll combine the two, by giving some examples of mathematical problems. They are drawn from all periods, to emphasize the continuity of mathematical thought. Any technical terms will be explained later.

(1) Is there a fraction whose square is exactly equal to 2?

(2) Can the general equation of the fifth degree be solved using radicals?

(3) What shape is the curve of quickest descent?

(4) Is the standard smooth structure on 4-dimensional space the only one possible?

(5) Is there an efficient computer program to find the prime factors of a given number?

Each problem requires some comment and preliminary explanation, and I'll consider them in turn.

The first was answered—negatively—by the ancient Greeks. The discovery is generally credited to the Pythagorean school in about 550 BC. It was an important question because the Greek geometers knew that the diagonal of the unit square has a length whose square is 2. Thus there exist natural geometric quantities that cannot be expressed as fractions (ratios of whole numbers). There is a legend that a hundred oxen were sacrificed to celebrate the discovery. It may well be false (no documents from the Pythagoreans survive), but the discovery certainly warranted such a celebration. It had a fundamental effect on the next 600 years of Greek mathematics, tilting the balance away from arithmetic and algebra towards geometry. Despite the marvellous discoveries of Greek geometry, this lack of balance seriously distorted the development of mathematics. Its repercussions were still being felt 2,000 years later, when Isaac Newton and Gottfried Leibniz were inventing the calculus.

The second problem came to prominence during the Renaissance. A *radical* is a formula involving only the usual operations of arithmetic (addition, subtraction, multiplication, and division), together with the extraction of roots. The *degree* of an equation is the highest power of the unknown occurring in it. Equations of degrees 1 and 2 (linear and quadratic) were effectively solved by the ancient Babylonians around 2000 BC. Degrees 3 and higher presented an unsolved problem of great notoriety, which withstood all attacks until the sixteenth century, when the cubic (degree 3) equation was solved by Scipione del Ferro of Bologna and Niccolo Fontanta (nicknamed Tartaglia, 'the stammerer') of Brescia. Soon after, Lodovico Ferrari solved the quartic (degree 4). In all cases the solution was by radicals. But the quintic equation (degree 5) resisted all efforts. In the nineteenth century two young mathematicians, Niels Henrik Abel and Évariste Galois, independently proved the impossibility of such a solution. The work of Galois led to much of modern algebra and is still a fertile field of research.

The third problem concerns (an idealized mathematical model of) the motion of a particle under gravity down a frictionless wire, from one point to a lower one (displaced some distance sideways). Depending on the shape of this wire, the particle will take differing times to make the journey. What shape of wire leads to the shortest trip? Galileo Galilei got the wrong answer in 1630 (he thought it was an arc of a circle). John Bernoulli posed the problem again, under the name 'brachistochrone', in 1696: solutions were found by Newton, Leibniz, Guillaume L'Hôpital, Bernoulli himself, and his brother James. The solution is an inverted cycloid (the

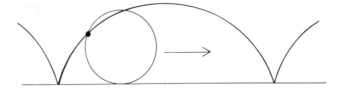

Fig. 1.   A point on the rim of a moving wheel traces out a cycloid.

path traced by a point on the rim of a moving wheel). This elegant problem, with its tidy and satisfyingly classical answer, led to the Calculus of Variations, which in short order revolutionized first mechanics, then optics. Today it remains a cornerstone of mathematical physics, with its influence being felt in Quantum Mechanics and General Relativity, as well as pure mathematics.

The fourth problem is of modern vintage, and belongs to an area known as Differential Topology. In this subject, one studies multidimensional analogues of surfaces, or *manifolds*, and these are equipped with a concept of 'smoothness'—so that, for example, you can decide whether a curve drawn on a manifold has a sharp corner or not. A major question concerns the *uniqueness* of the smooth structure on a given manifold. In 1963 John Milnor startled the mathematical world by finding more than one smooth structure on the 7-dimensional sphere. Generalizations of his *exotic spheres* have since been found in other dimensions. However, it was still expected that more homely manifolds, especially ordinary n-dimensional space, ought to possess only *one* smooth structure—the one everybody uses to do calculus. And this could be proved (not easily) for all dimensions except four. But in 1983 Simon Donaldson showed that there is an *exotic* smooth structure on 4-dimensional space, in addition to the standard one. In fact Donaldson solved a rather different problem, using some very new ideas born of an interaction between topology, algebra, and mathematical physics. When Donaldson's results are combined with the earlier work of Michael Freedman, Karen Uhlenbeck, and Clifford Taubes, the exotic 4-space emerges.

The fifth problem comes from a novel branch of mathematics known as Complexity Theory. The advent of the electronic computer has focused attention not just on ways to solve mathematical problems, but on the *efficiency* of their solutions. Donald Knuth has remarked that a major difference in style between mathematicians and computer scientists is that the former do not, on the whole, worry about the *cost* (in effort, time, or money) of a solution. In Complexity Theory, it is this cost (measured in terms of the total number of arithmetical operations) that is paramount. A basic distinction is the way the cost rises as the size of the input data grows. Typically it either grows slowly (slower than some fixed power of the input size) or very rapidly ('exponential growth', where the cost is multiplied by some definite amount for each single increase in input size). Recall that a *prime* number is one having no divisors, such as 11 or 29. Every number is a product of primes, but it is far from easy to find these factors. Trying all possible factors in turn is a procedure whose cost grows exponentially as the size of the number increases. The question is: can a more efficient method be found—one whose cost grows like a fixed power? Nobody knows the answer. The solution is important not only in Complexity Theory, but for cryptography: a number of 'theoretically unbreakable' military and diplomatic codes would cease to be unbreakable if the answer were 'yes'.

The historical thread

One feature illustrated by my selection of problems is the unusually long lifetime of mathematical ideas. The Babylonian solution of quadratic equations is as fresh and as useful now as it was 4,000 years ago. The calculus of variations first bore fruit in classical mechanics, yet survived the quantum revolution unscathed. The *way* it was used changed, but the mathematical basis did not. Galois's ideas remain at the forefront of mathematical research. Who knows where Donaldson's may lead? Mathematicians are generally more aware of the historical origins of their ideas than many other scientists are. This is not because nothing important has happened in mathematics recently—quite the reverse is true. It is because mathematical ideas have a permanence that physical theories lack.

Really good mathematical ideas are hard to come by. They

result from the combined work of many people over long periods of time. Their discovery involves wrong turnings and intellectual dead ends. They cannot be produced at will: truly novel mathematics is not amenable to an industrial 'Research and Development' approach. But they pay for all that effort by their durability and versatility. Ptolemy's theory of the solar system is of historical interest to a modern cosmologist, but he doesn't *use* it in serious research. In contrast, mathematical ideas thousands of years old are used every day in the most modern mathematics, indeed in all branches of science. The cycloid was a fascinating curiosity to the Greeks, but they couldn't *do* anything with it. As the brachistochrone, it sparked off the calculus of variations. Christiaan Huygens used it to design an accurate clock. Engineers today use it to design gears. It shows up in celestial mechanics and particle accelerators. That's quite a career for one of such a humble upbringing.

Yes, but look at it this way . . .

As well as problems and examples, we mustn't ignore a third source of mathematical inspiration: the search for the 'right context' for a theorem or an idea. Mathematicians don't just want to get 'the answer' to a problem. They want a method that makes that answer seem inevitable, something that tells them what's *really* going on. This is something of an infuriating habit, and can easily be misunderstood; but it has proved its worth time and time again.

For example, Descartes showed that by introducing coordinates into the plane, every problem in geometry can be reformulated as an equivalent problem in algebra. Instead of finding where curves meet, you solve simultaneous equations.

It's a magnificent idea, one of the biggest mathematical advances ever made. But let me 'prove' to you that it's totally worthless. The reason is simple: there is a routine procedure whereby *any* algebraic calculation can be transformed into an equivalent geometric theorem, so anything you can do by algebra, you could have done by good old geometry. Descartes's reformulation of geometry adds precisely nothing. See? Like I said, it's useless.

In point of fact, however, the reformulation adds a great

deal—namely, a new viewpoint, one in which certain ideas are much more natural than they would have been in the original context. Manipulations that make excellent sense to an algebraist can appear very odd to a geometer when recast as a series of geometric constructions; and the same goes for geometrical ideas reformulated as algebra. It's not just the *existence* of a proof that is important to mathematicians: it's the flow of ideas that lets you think the proof up in the first place. A new viewpoint can have a profound psychological effect, opening up entirely new lines of attack. Yes, *after* the event the new ideas can be reconstructed in terms of the old ones; but if we'd stuck to the old approach, we'd never have thought of them at all, so there'd be nothing to reconstruct *from*.

This point is often misunderstood, even in today's enlightened atmosphere. Critics of 'non-standard analysis'—which involves genuine infinitesimals (see chapter 8)—typically object that every theorem that can be proved from the non-standard point of view can also be proved using traditional 'standard' analysis. They're right; and they've missed the point. The point, again, is not the existence of a proof within a given framework: it is whether that proof has a natural flow of ideas, whether it might actually *occur* to somebody.

By exploiting Descartes's connection between geometry and algebra, it is relatively straightforward to prove that angles cannot be trisected using ruler and compasses. This problem remained unsolved from at least the time of the later Greek commentators, say around AD 600, until 1837, when Pierre Wantzel gave an algebraic proof of the impossibility of trisection. In *principle* you could rewrite each step in Wantzel's proof as a geometric theorem—so why didn't the Greeks think of it? Or anybody else for two thousand years and more? Because in *practice*, the only way anyone would ever find the geometric line of proof would be to start from Wantzel's algebra.

As I said, people often misunderstand this restless search for new formulations, new viewpoints, new ways of seeing old results and ideas. Mathematicians are often accused of wanting to reformulate problems from different points of view in order to put off the evil hour when they actually have to try to *solve* them. It's true that there's no point in reformulating a problem if, at the end of the day, you learn nothing new about it. But it's important not to expect major new results straight away. I call this kind of thinking the 'Rolls-Royce Problem' or the 'Dog-in-the-Manger Effect'. Suppose tomorrow you come up with a really good idea for improving the design of cars—say, using alcohol as a fuel, or installing an electric motor. Opponents of your approach may then demand that you prove its superiority by manufacturing—immediately—a better car than a Rolls-Royce. Of course you can't—even if your idea really is better. Half a century or more of engineering has gone into the design of a Rolls-Royce; they're expecting you to do better, overnight, with no funding. The world is full of obsolete products for this very reason—the state of the art in Mexican taxis is a Volkswagen beetle; manufacturing industry is still hooked on the computer language FORTRAN. But propose a novel reformulation of somebody else's research area—especially if it uses ideas that they've not been trained in—and ten to one they'll challenge you to justify it by building a mathematical Rolls-Royce. We'll see several case histories of this as we go: I'll leave you the pleasure of spotting them.

Despite all this, fashions do change, formulations do get revamped—and Rolls-Royces do get built overnight, sometimes. After all, this is mathematics, not manufacturing, and the only limitations are those of human imagination.

The mainstream

The Mississippi river is 2,348 miles long, and drains an area of 1.24 million square miles. In many places it seems just a maze of twists and turns and dead-end backwaters, yet it manages to flow virtually the entire length of the continent of North America, from Lake Itasca in Minnesota to New Orleans in Louisiana. There are tributaries and meanders and ox-bow lakes cut off entirely from the main river; but the main river is there, and it can be picked out by anyone able to gauge the strength of the current. At its mouth, the Mississippi enters the Gulf of Mexico in an enormous muddy delta, with its own side-shoots and main channels, whose shape and interconnections are constantly changing.

Mathematics is much like the Mississippi. There are side-shoots and dead ends and minor tributaries; but the mainstream is there, and you can find it where the current—the mathematical power—is strongest. Its delta is research mathematics: it is growing, it is going somewhere (but it may not always be apparent where), and what today looks like a major channel may tomorrow clog up with silt and be abandoned. Meanwhile a minor trickle may suddenly open out into a roaring torrent. The best mathematics always enriches the mainstream, sometimes by diverting it in an entirely new direction.

The golden age

Until quite recently, most mathematics taught in schools belonged to the 'classical' period, the most recent part being the calculus (dating from about 1700). Even with the advent of so-called 'modern mathematics', most material taught is at least a century old! Partly in consequence of this, mathematics is thought by many to be a dead subject, and it often comes as a revelation that anything new remains to be discovered.

Can we place in history the Golden Age of Mathematics? Was it the time of Euclid, when the logical foundations of geometrical reasoning were laid? The development of the Hindu-Arabic number system? The Renaissance flowering of algebra? Descartes's coordinate geometry? The calculus of Newton and Leibniz? The vast edifice of eighteenth-century natural philosophy, bending mathematics to the needs of astronomy, hydrodynamics, mechanics, elasticity, heat, light, and sound? If there has indeed been such a thing as a golden age of mathematics, it is none of these. It is the present. During the past fifty years, more mathematics has been created than in all previous ages put together. There are more than 1,500 mathematical research journals, publishing some 25,000 articles every year (in over a hundred languages). In 1868 The *Jahrbuch über die Fortschritte der Mathematik* (*Yearbook of Mathematical Research*) listed just twelve categories of mathematical activity; in *Mathematical Reviews* for 1992 there are more than sixty. Of course it is quality, not quantity, that we should use as our yardstick; but by this standard too, the golden age is the modern one.

Some observers have professed to detect, in the variety and freedom of today's mathematics, symptoms of decadence and decline. They tell us that mathematics has fragmented into unrelated specialities, has lost its sense of unity, and has no idea where it is going. They speak of a 'crisis' in mathematics, as if the whole subject has collectively taken a wrong turning. There is no crisis. Today's mathematics is healthy, vigorous, unified, and as relevant to the rest of human culture as it ever was. Mathematicians know very well where they think they and their subject are going. If there appears to be a crisis, it is because the subject has become too large for any single person to grasp. This makes it hard to get an effective overview of what is being done, and what it is good for. To an outside observer, the mathematician's love of abstract theory and hair-splitting logic can appear introverted and incestuous. His apparent lack of attention to the real world suggests the self-centred complacency of the ivory tower. But today's mathematics is not some outlandish aberration: it is a natural continuation of the mathematical mainstream. It is abstract and general, and rigorously logical, not out of perversity, but because this appears to be the only way to get the job done properly. It contains numerous specialities, like most sciences nowadays, because it has flourished and grown. Today's mathematics has succeeded in solving problems that baffled the greatest minds of past centuries. Its most abstract theories are currently finding new applications to fundamental questions in physics, chemistry, biology, computing, and engineering. Is this decadence and decline? I doubt it.

2

The price of primality

Suppose that the cleaning lady gives p and q by mistake to the garbage collector, but that the product pq is saved. How to recover p and q? It must be felt as a defeat for mathematics that the most promising approaches are searching the garbage dump and applying memo-hypnotic techniques.

Hendrik W. Lenstra Jr.

Some of the best problems in mathematics are also the simplest. It is difficult to think of anything simpler and more natural than arithmetic. The additive structure of the whole numbers 1, 2, 3, . . . is a bit too simple to hold any great mysteries, but the multiplicative structure poses problems that still inspire creative work after thousands of years. For instance, take the very simple and natural idea of a prime number—one that cannot be obtained by multiplying two smaller numbers. It was known from ancient times, as an empirical fact, that all numbers are products of primes in a unique way; and the Greeks managed to prove it. The problem immediately arises of *finding* these primes for any given number. It isn't hard to think of 'theoretical' methods. For example, the Greeks developed one called the sieve of Eratosthenes, which boils down to working out all possible products or primes and checking whether the number you want occurs anywhere. It's a bit like building a house by systematically trying all possible ways to arrange ten thousand bricks and then choosing the one that seems most habitable, and for practical purposes it's hopelessly inefficient. Even today, there remains a great deal of room for improvement, and basic questions are still unanswered.

The source of the difficulty lies in the schizophrenic character of the primes. Don Zagier, an expert number theorist, describes it like this:

There are two facts about the prime numbers of which I hope to convince you so overwhelmingly that they will be permanently engraved in your hearts. The first is that they are the most arbitrary objects studied by mathematicians: they grow like weeds among the natural numbers, seeming to obey no other law than that of chance, and nobody can predict where the next one will sprout. The second fact states just the opposite: that the prime numbers exhibit stunning regularity, and there are laws governing their behaviour, and that they obey these laws with almost military precision.

Zagier is referring in particular to the extensive theories developed during the past century on the distribution of prime numbers. For example there is the Prime Number Theorem, guessed on numerical evidence by Carl Freidrich Gauss in 1792 and proved independently by Jacques Hadamard and Charles-Jean de la Vallée Poussin in 1896. This states that for large x the number of primes less than x is approximated ever more closely by $x/\log x$. But in this chapter we confine attention to less lofty matters, concentrating on the computational aspects of the primes. Even with these simple raw materials, mathematics has tailored a stunning garment. Almost as a side issue, we observe that number theory, long considered the most abstract and impractical branch of mathematics, has acquired such a significance for cryptography that there have been serious attempts to have some of its results classified as military secrets.

Divide and conquer

The sequence of primes begins 2, 3, 5, 7, 11, 13, 17, 19, 23, 29, . . . and extends, in rather irregular fashion, beyond reasonable limits of computation. The largest *known* prime is $2^{756839} - 1$, found by scientists at AEA Harwell, near Oxford, in 1992. But we can be certain that even larger primes exist, even if we don't know what they are, because Euclid proved that there are infinitely many primes. The gist of the proof is to assume that there are only finitely many primes, multiply the lot together, and add 1. The result must have a prime factor, but this can't be any of the original list, since these all leave remainder 1. This is absurd, therefore no finite list existed to begin with. So if we form the enormous number $2 \cdot 3 \cdot 4 \cdot 5 \ldots (2^{756839} - 1) + 1$ we know that *all* of its prime factors must be larger than Harwell's gigantic prime. However, we have no practical way to find any of these factors, and thereby hangs a tale.

Primality testing

The two central practical problems in prime number theory are to:

(1) Find an efficient way to decide whether a given number is prime.

(2) Find an efficient way to resolve a given number into prime factors.

Clearly (2) is at least as hard as (1), since any solution to (2) will yield (1) as a special case. The converse is not obvious (and probably not true): we shall see below that it is possible to prove that a number is not prime, without actually exhibiting any factor explicitly. 'Efficient' here implies that the length of the calculation should not grow too rapidly with the size of the number. I'll have a lot more to say about *that* idea in Chapter 21.

The most direct way to find a factor of a number, or prove it prime if no factor exists, is *trial division*: divide by each number in turn (up to its square root) and see if anything goes exactly. This is *not* efficient, however. Imagine a computer capable of performing a million trials per second. It will take about a day to deal with a 30-digit number; a million years for a 40-digit number; and ten quadrillion years for a 50-digit number—longer than the current predictions for the age of the universe! Obviously there may be short cuts; but the fastest method known takes about a month for a 120-digit number, and a century or more for a 130-digit number. Until very recently, both problems were in a very sorry state. But in the last five years, enormous progress has been made on (1), primality testing. A 100-digit number can be proved prime (or not) in about 15 seconds on a fast computer; a 200-digit number takes 2 minutes. A 1000-digit number might take a day or two. In contrast, to solve (2) for a 200-digit number still looks utterly impracticable.

Finite arithmetic

To understand how this progress has come about, we must start with an idea first codified by Gauss in his *Disquisitiones Arithmeticae*, a milestone in number theory. Consider a clock, numbered (in unorthodox fashion) with the hours 0, 1, 2, . . . , 11. Such a clock has its own peculiar arithmetic. For example, since three hours after 5 o'clock is 8 o'clock, we could say that 3 + 5 = 8, as usual. But 3 hours after 10 o'clock is 1 o'clock, and 3 hours after 11 o'clock is 2 o'clock; so by the same token, 3 + 10 = 1 and 3 + 11 = 2. Not so standard! Nevertheless, this 'clock arithmetic' has a great deal going for it, including almost all of the usual laws of algebra. Following Gauss, we describe it as arithmetic to the *modulus* 12, and replace '=' by the symbol '≡' as a reminder that some monkey-business is going on. The relation '≡' is called *congruence*. In arithmetic *modulo* (that is, to the modulus) 12, all multiples of 12 are ignored. So 10 + 3 = 13 ≡ 1 since 13 = 12 + 1 and we may ignore the 12.

Any other number n may be used as the modulus: now multiples of n are neglected. The resulting arithmetical system consists only of the numbers 0, 1, 2, . . . , $n - 1$, and has its own multiplication and subtraction, as well as addition. Division is less pleasant; but if n is a *prime* then it is possible to divide by any non-zero number. For example, modulo 7 we find that $3/5 \equiv 2$, since $2.5 = 10 \equiv 3$. This bizarre arithmetic was introduced by Gauss because it is ideal for handling questions of *divisibility*. Number theory has used it for this purpose ever since.

Fermat's little theorem

Two important theorems in mathematics bear the name of Pierre de Fermat, a French lawyer and amateur number theorist *par excellence*. One, Fermat's Last Theorem, isn't a real theorem at all, but a major open problem (see chapter 3). The other, often called the 'little theorem', is so simple that a proof is only a few lines long; yet it is a masterpiece of ingenuity and a result with astonishing implications.

It is this. Suppose p is a prime and a is any number not divisible by p. Then $a^{p-1} \equiv 1 \pmod{p}$. The 'mod p' just reminds us that p is the modulus. In other words, a^{p-1} leaves remainder 1 on division by p. For example, we expect a remainder 1 on dividing $2^{10} = 1024$ by 11, since 11 is prime. Now $1024 = 93 \cdot 11 + 1$, so this is correct. On the other hand, $2^{11} = 2048 = 170 \cdot 12 + 8$, so leaves remainder 8 on division by 12. We have thus succeeded in proving that 12 is not prime, *without* exhibiting any specific factors. This fact is hardly novel, but similar calculations can be made for very large values of the number p, where the outcome is less obvious. Calculating $a^{p-1} \pmod{p}$ can be done efficiently, in a time that grows like the *cube* of the number of digits in p. So here we have an efficient test for non-primality—provided we know which value of a to choose!

Trapdoor codes

Why should anyone be interested in large primes, and factorization? Aside from the sheer joy of it all, there is a practical reason, which I hinted at earlier: applications to cryptography. A great deal of military effort has gone into the search for an 'unbreakable' but effective code. In the late 1970s Ralph Merkle, Whitfield Diffie, and Martin Hellman proposed a new kind of code, called a public-key cryptosystem. In using any code, there are two basic steps: *encoding* a message, and *decoding* it again. For most codes, anyone who knows how to perform the first step also knows how to perform the second, and it would be unthinkable to release to the enemy (or the public, which many governments seem to view in the same light) the method whereby a message can be turned into code. Merely by 'undoing' the encoding procedures, the enemy would be able to break all subsequent coded messages.

Merkle, Diffie, and Hellman realized that this argument is not totally watertight. The weasel word is 'merely'. Suppose that the encoding procedure is very hard to undo (like boiling an egg). Then it does no harm to release its details. This led them to the idea of a *trapdoor function*. Any code takes a message M and produces a coded form $f(M)$. Decoding the message is tantamount to finding an inverse function f^{-1}, such that $f^{-1}(f(M)) = M$. We call f a trapdoor function if f is very easy to compute, but f^{-1} is very hard, indeed impossible for practical purposes. A trapdoor function in this sense isn't a very practical code, because the legitimate user finds it just as hard to decode the message as the enemy does. The final twist is to define f in such a way that a single extra piece of (secret) information makes the computation of f^{-1} easy. This is the only bit of information you can't tell the enemy.

In the RSA system, designed by Ted Rivest, Adi Shamir, and Leonard Adleman, this is achieved with the help of Fermat's little theorem. Begin with two large primes p and q, which are kept secret; but release to the public their product $n = pq$ and a further number E, the *encoding key*. It is easy to represent any message by a sequence of digits; break this into blocks which, thought of as numbers, are less than n. Then it is enough to encode block by block. To do this, change each block B into the number $C \equiv B^E \pmod{n}$. Then C is the coded message. To decode, you need to know a *decoding key* D, chosen so that DE is congruent to 1 modulo $(p - 1)(q - 1)$. Then a slight generalization of Fermat's little theorem, due to Leonhard Euler, shows that $C^D \equiv B \pmod{n}$, so you can recover B from C by a similar process.

The point is that while Joe Public knows n and E, he doesn't know p and q, so he can't work out $(p - 1)(q - 1)$ and thereby find D. The designer of the code, on the other hand, knows p and q because those are what he started from. So does any legitimate receiver: the designer will have told him. The security of this system depends on exactly one thing: the notorious difficulty of factorizing a given number n into primes, alluded to in Lenstra's remark about the cleaning-lady at the head of this chapter. You don't believe him? OK, here's what the cleaning-lady left: $n = 2^{67} - 1$. If you can work out p and q *by hand* in less than three years of Sundays, you'll beat F. N. Cole, who first did it in 1903. (See the end of this chapter.)

Pseudoprimes

With this application in mind we return to the two theoretical problems: primality testing and factorization. The Fermat test for non-primality, mentioned earlier, is not foolproof. For each value of a there exist infinitely many non-primes p for which $a^{p-1} \equiv 1 \pmod{p}$. These p are called *pseudoprimes* to base a. Worse, there are non-primes p that are pseudoprimes to *all* bases a (not having a factor in common with p). Such p are called *Carmichael numbers*: the smallest is $561 = 3 \cdot 11 \cdot 17$. (It works because $3 - 1$, $11 - 1$, and $17 - 1$ all divide $561 - 1$, a pattern typical of Carmichael numbers. Try to see how Fermat's Little Theorem can be used to explain this.)

There is a related, but more complicated idea, due to G. L. Miller, which I won't describe in detail. It has the same feature that any number that fails the Miller test is known not to be prime, but passing the test is not of itself a guarantee of primality. A number that passes the Miller test is called a *strong pseudoprime* to base a. If not, then a is called a *witness* for (the non-primality of) p. It turns out that there is no analogue of Carmichael numbers for strong pseudoprimality. In fact most non-primes have a very small witness. For example, John Selfridge and S. S. Wagstaff found by direct computation that the only non-prime number less than 25 billion that does not have one of 2, 3, 5, and 7 as witness is 3215031751. But then 11 *is* a witness. By exploiting this we can obtain an efficient primality test for 9-digit numbers which will run on a small programmable calculator within two minutes.

Miller took this idea further. If we can show that every non-prime number has a sufficiently small witness, then we have an efficient test for primality which works for *all* numbers. He was able to establish that any odd non-prime n has a witness less than $70(\log n)^2$; but only by assuming one of the most notorious unproved conjectures in the whole of mathematics, the Riemann Hypothesis (see chapter 12). This puts us in a very curious position. We have a primality test, which we can be pretty certain is efficient; but we can't prove that it is. If in fact the Riemann Hypothesis is *true*—as most experts think likely—then the test really *is* efficient. As a practical matter, it presumably works efficiently on almost all numbers. An engineer, for instance, might well consider the problem to be solved. (I mean no offence here—in engineering the criterion 'it works' is arguably the only sensible test anyway.) Mathematicians find the situation totally unsatisfactory—I suppose because, although they think they have 'the answer', there is a gaping hole in their understanding of it. This isn't a point a mathematician is likely to slide over, either, because experience has shown that whenever a gap isn't understood, all sorts of beautiful and important ideas are lurking within it.

The Adleman–Rumely test

In 1980 Adleman and Robert Rumely *found* some of those lurking ideas, allowing them to modify the Miller test so that it would give a *guarantee* of primality or non-primality, with no probabilistic or conjectural fudging. The running-time has to be a little longer, but the method (especially as improved by Hendrik Lenstra soon afterwards) is entirely practicable for, say, 200-digit primes. Their idea is to extract extra information from a strong pseudoprime test, other than just 'pass' or 'fail', by using some ideas from algebraic number theory related to the so-called 'higher reciprocity laws'. Some very sophisticated mathematics, both classical and modern, is getting into the act, despite the relatively mundane problem that it is being applied to. For a k-digit number, their method has running time approximately $k^{\log \log k}$. This is 'nearly polynomial time' and makes the method a practical one. The exact bound on the running-time was conjectured by Adleman and Rumely, and proved soon after by Andrew Odlyzko and Carl Pomerance.

Schrödinger's cat

A very recent development in cryptography has nothing whatever to do with primes, but it's so nice that I can't bring myself to miss it out. Public key cryptosystems permit the methods for encrypting messages to be made public, with decryption being virtually impossible in the absence of a secret key. An emerging branch of the subject—quantum cryptography—goes much further. Charles C. Bennett, François Bessette, Gilles Brassard, Louis Salvail, John Smolin, and Artur Ekert—the first four in joint work, the last somewhat independently—have recently explained how ideas based upon the Einstein–Podolsky–Rosen 'paradox' in quantum theory can be used to guarantee secure distribution of a crytographic key.

In fact, quantum cryptography began with unpublished work of Stephen Wiesner in the late sixties, who explained how quantum effects can in principle be used to manufacture bank-notes immune to counterfeiting, and to transmit information securely. His work was eventually published in 1983 at

Bennett's instigation. The unfakeable banknote requires the ability to store a single polarized photon for several days without change of state. As Bennett and his co-authors say: 'Unfortunately the impact of the CRYPTO 82 conference had left most people under the impression that everything having to do with quantum cryptography was doomed from the start to being unrealistic.' However, in 1982 Bennett and Brassard realized that photons are not for *storing* information, but for *transmitting* it. Quantum versions of many mainstays of cryptography then appeared, including the self-winding reusable one-time pad, quantum coin-tossing, quantum oblivious transfer, zero-knowledge protocols, and secure two-party computations. Here I'll consider only the quantum key distribution channel, also originally invented by Bennett and Brassard.

Security of a cryptotext used to depend upon keeping the entire process of encryption and decryption secret. As we have seen, the encryption and decryption procedures can be public knowledge, provided only that a specific piece of information, the *key*, is kept secret from illegitimate users or eavesdroppers. The key can be any random string of digits. The danger with this approach is that all eggs are invested in one basket, the key. Before secure communication can begin, the key must be distributed *securely* to users. During World War II one rather amateur group of spies distributed new code keys by radio to possibly compromised agents, using the old (equally compromised) code. This is hardly an acceptable solution to the problem; but more subtle distribution methods might also be open to eavesdropping.

Consider a specific example, an eavesdropper trying to monitor a communication channel, say by tapping a telephone line. Observations of any system disturb it and so leave traces. Legitimate users can try to guard against this by making their own measurements on the line, to detect the effects of tapping. However, the tappers will be safe provided they can make measurements that produce smaller disturbances than the legitimate users can detect. In classical (that is, non-quantum) mechanics, this kind of 'arms race' of increasingly sensitive measurements can go on indefinitely. ('Big bugs have little bugs', et cetera.) However, in quantum mechanics it cannot, because of the Heisenberg uncertainty principle.

Measurements of some quantities, however delicately conducted, may change other related quantities by significant and irreducible amounts. In the conventional interpretation, a quantum system exists in a superposition of possible states, *until* a measurement is made, 'forcing' it into a pure state. This view is dramatized by the unenviable fate of Schrödinger's cat, which lives inside an impenetrable box, which is equipped with a device that ensures that the cat is alive if a particular particle has positive spin, dead if it has negative spin. Until the box is opened, the unfortunate beast is neither alive nor dead, but— adherents of the 'Copenhagen interpretation' of quantum mechanics maintain—in a quantum superposition of the two states.

If I may digress for a moment: this interpretation may well be wrong. Recent experiments suggest that the cat will be in a good old classical state, either alive or dead—but you won't know which until you open the box. This is *not* the same as a quantum superposition, since, as T. S. Eliot put it, 'THE CAT HIMSELF KNOWS, and will never confess'. The device that converts particle spin into condition-of-cat itself makes a measurement within the meaning of the act. It is the amplification of quantum information into a large-scale effect that constitutes 'measurement', not necessarily the action of a human mind. How this really happens remains mysterious. This possibility notwithstanding, I'll stick to the Copenhagen interpretation, because it lets me use the next section heading:

Schrödinger's cat-flap

At this point the EPR paradox, proposed in 1935 by Albert Einstein, Boris Podolsky, and Nathan Rosen, enters the story. Their ideas were simplified by David Bohm, and it is his version that I'm going to describe. The proton is a particle with spin $\frac{1}{2}$. Its (spin) angular momentum, measured in any direction, is therefore $\pm\frac{1}{2}\hbar$, where $\hbar = h/2\pi$ and h is Planck's constant. Begin with a system of two protons, very close together, of total angular momentum zero, a situation that can be realised in practice. If the angular momentum of one proton in a given direction is $\frac{1}{2}\hbar$, then that of the other must be $-\frac{1}{2}\hbar$; if that of the first is $-\frac{1}{2}\hbar$ then that of the second must be $\frac{1}{2}\hbar$. It is not necessary to measure these quantities for this relation to hold—indeed it is important not to, since measurements in quantum mechanics disturb the system being measured. Assume that the protons move apart until the distance between them is large: the relations between the angular momenta will remain valid. If we now measure a component of angular momentum for one proton, then the result forces a definite state of angular momentum in that direction for its distant companion. This kind of 'action at a distance' seemed to Einstein, Podolsky, and Rosen to be paradoxical, and they concluded that quantum mechanics must be an incomplete description of reality. However, John Bell subsequently derived a theoretical relationship, Bell's inequality, that permits experiments to test the EPR paradox. The results confirmed the predictions of quantum mechanics.

The quantum key distribution channel repeats the EPR scenario once per bit (binary digit's worth) of transmitted message: it is a source that emits pairs of spin $\frac{1}{2}$ particles in opposite directions. That is, a box with a cat-flap at each end, which emits pairs of perfectly anticorrelated Schrödinger's cats in a constant stream. If one cat in a given pair is alive, then the other is dead, but until their state is measured, we do not know which is which. At opposite ends of this axis are two time-honoured characters in the lore of cryptography who represent the legitimate users of the channel, known as Alice and Bob. Their task is to extract from the sequence of EPR-linked particles a common key, and also to monitor the channel against eavesdropping.

To this end, they each measure components of angular momentum, in orientations that they select randomly and independently for each incoming particle, thereby determining the conditions of their respective cats. Alice has the choice of measuring at angles of 0°, 45°, or 90° in a plane perpendicular to the communication axis; Bob at angles of 45°, 90°, or 135°. Each such measurement yields a result of $+1$ or -1 (in units of $\frac{1}{2}\hbar$), and potentially reveals one bit of information. If Alice and Bob measure angular momentum in identical orientations, then their results are totally anticorrelated: the state of Bob's cat is always the exact opposite to that of Alice's. However, if they use different orientations, it can be shown that a particular combination S of correlations must be $-2\sqrt{2}$.

After transmission, Alice and Bob publicly announce the sequence of orientations they have chosen, and thus can privately divide their measurements into two groups: those for which they used the same orientation, and those for which they used different ones. They also announce the actual measurements they obtained in the second group, so that they both can compute the quantity S. If the channel has not been disturbed by an eavesdropper, the value should be $S = -2\sqrt{2}$. If that is the case, then the first group of perfectly anticorrelated measurements (whose values are known only to Alice and Bob) can be used to establish a common key.

Experimental methods designed to test Bell's inequalities could be used to implement this scheme (over short distances) in a relatively practical manner. For quantum cryptography, Schrödinger's cat is out of the bag.

Cole's Answer
$2^{67} - 1 = 193707721 \times 761838257287$.

3

Marginal interest

In passing, may I request any reader of this
section who imagines he has a proof not to send
it to me? I have examined well over a hundred
fallacious attempts, and I feel that I have done
my share. One such, many years ago, stuck me
for three weeks. I felt that there was a mistake,
but couldn't find it. In desperation I turned the
author's manuscript over to a very bright girl
in my trigonometry class, who detected the
blunder in half an hour.

Anyone contemplating a proof may be
interested in what Hilbert said in 1920 when
asked why he did not try: 'Before beginning I
should put in three years of intensive study,
and I haven't that much time to squander on a
probable failure.'

Eric Temple Bell

One of the greatest number theorists of the seventeenth century was the lawyer Pierre de Fermat. His fame rests on his
correspondence with other mathematicians, for he published
very little. He would set challenges in number theory based on
his own calculations; and at his death he left a number of
theorems whose proofs were known, if at all, only to himself.
The most notorious of these was a marginal note in his own
copy of the *Arithmetica* of Diophantus: 'To resolve a cube into
the sum of two cubes, a fourth power into fourth powers, or in
general any power higher than the second into two of the same
kind, is impossible; of which fact I have found a remarkable
proof. The margin is too small to contain it.' If Fermat really
did have a proof, nobody has the slightest idea what it was.
What we currently know about Fermat's Last Theorem, as it
has come to be called, requires methods that could not poss-

ibly have been available in the seventeenth century. But whether Fermat had noticed something that has eluded everybody since, or whether he was deluding himself, his almost casual remark has been responsible for a vast amount of mathematics. Fermat's Last Theorem is an example of a problem so good that even its failures have enriched mathematics beyond measure.

Alexandrian algebra

In 332 BC Alexander the Great founded a city in Egypt, and modestly called it Alexandria. But he died in 323 BC, before the city had been completed. In the ensuing political instability Alexander's empire split into three, one part being Egypt under the Ptolemaic dynasty. The main mathematical activity after the classical Greek period was within the Ptolemaic empire, especially at Alexandria. Science and mathematics flourished there until the destruction of the Alexandrian library, begun by the Romans under Theodosius in AD 392 and completed by the Moslems in 640—'if it's in the Koran, it's superfluous; if not, then it's heresy'. It is estimated that under Theodosius over 300,000 manuscripts were destroyed.

A century or so before its final decline, Alexandria produced the high point of Greek algebra, the work of Diophantus. Very little is known about Diophantus—for example, it is only an assumption that he was Greek. He wrote a number of books, of which the *Arithmetica* was the most important. One might describe him as the Euclid of algebra. He introduced a symbolic notation with different symbols for the square, cube, and so forth of the unknown. And he wrote about the solution of equations. By a solution he meant a rational number, usually an integer. This was not so much an explicit requirement as a tacit assumption: those were the only kinds of number that could be handled by *arithmetic*. Today we use the term *Diophantine equation* for one whose solutions are required to be integers.

Among the problems treated by Diophantus is that of Pythagorean triples: whole numbers that form the sides of a right triangle. It was known from very ancient times that a triangle whose sides are 3, 4, and 5 units will have a right angle. Using Pythagoras' Theorem, the general problem boils down to finding integers a, b, c such that $a^2 + b^2 = c^2$. The Old Babylonian tablet Plimpton 322, dating from about 1900–1600 BC, lists fifteen such triples. Diophantus tackles the general problem—though using special cases to exemplify the method. For example, Problem 8 in Book 1 asks for a division of 16 into two squares, and obtains the answer 256/25 and 144/25. In modern notation, the general solution is given by

$$a = k(u^2 - v^2), \quad b = 2kuv, \quad c = k(u^2 + v^2)$$

for arbitrary integers k, u, v. Thus there are infinitely many solutions, and they may be given in terms of polynomial functions of three parameters. Diophantus studied many other kinds of equation, for example making a cubic expression equal to a square. His aim was to find *some* solution rather than all solutions. As Morris Kline says: 'He has no *general methods*. His variety of methods for the separate problems dazzles rather than delights. He was a shrewd and clever virtuoso but apparently not deep enough to see the essence of his methods and thereby attain generality.' To some extent the same problem has plagued the theory of Diophantine equations for most of its history; but in the last two decades there has been enormous progress, and a great deal of unity and order is at last emerging.

The professional amateur

Fermat, the son of a tradesman, trained as a lawyer in Toulouse in the early 1600s. He created, almost singlehandedly, what is now called the Theory of Numbers (or Number Theory): the mathematics of whole numbers. In his book *The Mathematics of Great Amateurs* Julian Coolidge refuses to include Fermat on the grounds that he was so good he should be considered a professional. Fermat certainly had a powerful mathematical brain, and he didn't confine it to number theory. Some of his work anticipated basic ideas of calculus and probability. Among his lasting results is one stating that every prime of the form $4n + 1$ is a sum of two squares. For example $17 = 4^2 + 1^2$, $137 = 11^2 + 4^2$. Few of Fermat's proofs survive, but we know he didn't just guess his results, because his letters to other mathematicians occasionally include details of proofs. This particular result Fermat proved by a method of his own invention, called *infinite descent*. The basic idea is to show that if the theorem does not hold for some prime of the form $4n + 1$, then it also fails for some smaller prime of that form. Descending indefinitely we see that it must fail for the smallest such prime, namely 5. But $5 = 2^2 + 1^2$, a contradiction. To modern eyes this is just a variant of the method of mathematical induction, but in Fermat's time it was original.

While pondering Diophantus's work on Pythagorean triples, Fermat must have started thinking about the analogous problem for cubes, fourth powers, and so on: that is, the *Fermat Equation* $a^n + b^n = c^n$ ($n \geq 3$). We know this because of the aforementioned marginal note, which asserts that there are *no* solutions in non-zero integers, for $n \geq 3$. It is easy to see that it suffices to prove this for $n = 4$ and n any odd prime. A sketch of Fermat's proof for $n = 4$ is known. Euler obtained a proof when $n = 3$. The case $n = 5$ was proved by Peter Lejeune Dirichlet in 1828 and Adrien-Marie Legendre in 1830. In 1832 Dirichlet proved it for $n = 14$: it is curious that this is easier to deal with than the corresponding prime $n = 7$, and others tried to repair the omission. In 1839 Gabriel Lamé offered a proof for $n = 7$ but there were errors. Gauss had a go, failed, and wrote to Heinrich Olbers that the problem 'has little interest for me, since a multitude of such propositions, which one can neither prove nor refute, can easily be formulated'. Even the greatest mathematician suffers from the occasional attack of sour grapes.

Algebraic numbers

The key to further progress is the idea of an *algebraic number*—one satisfying a polynomial equation with rational coefficients. For example $\sqrt{2}$ is algebraic since it satisfies the equation $(\sqrt{2})^2 - 2 = 0$. (Some numbers, such as π, are not algebraic: these are called *transcendental*. See chapter 4 for further discussion.) It turns out that algebraic numbers have their own kind of arithmetic, with satisfactory generalizations of integers, primes, and so on. For instance, in the system of *Gaussian integers* $a + b\sqrt{-1}$, where a and b are ordinary integers, every number is a unique product of primes. More generally, *algebraic integers* are numbers that satisfy a polynomial equation with integer coefficients and leading coefficient

1. Sums and products of algebraic integers are again algebraic integers. The main motivation for introducing algebraic numbers into number theory is that they provide information on Diophantine equations. For example, Fermat stated that the only integer solutions of the equation $y^2 + 2 = x^3$ are $y = \pm 5$, $x = 3$. Here's a sketch proof using algebraic integers of the form $a + b\sqrt{-2}$. We bring these into the problem because we can factorize the left-hand side, to get

$$(y + \sqrt{-2})(y - \sqrt{-2}) = x^3.$$

By a bit of fiddling we find that we can assume the two numbers on the left have no common factor. Now, in ordinary integers, if the product of two numbers without common factors is a cube, then each separately is a cube. So, assuming this result is still true for algebraic integers, we have in particular that

$$y + \sqrt{-2} = (a + b\sqrt{-2})^3.$$

Expanding and comparing coefficients of $\sqrt{-2}$ we see that

$$1 = b(3a^2 - 2b^2)$$

whose only solutions are easily found by trial and error. We must have $b = \pm 1$, and $3a^2 - 2b^2 = \pm 1$. So $a = \pm 1$ and $b = 1$. Then $x = 3$ and $y = \pm 5$ as claimed. As it stands this isn't a rigorous proof, because we can't be sure that the result about cubes is still true for integers of the form $a + b\sqrt{-2}$. In fact this can be proved, as a consequence of unique prime factorization of such integers. The proof isn't totally obvious, but it's not especially deep.

This is an excellent example of the way in which the introduction of a new gadget—here the numbers $a + b\sqrt{-2}$—lets us deduce results about more ordinary objects of mathematics, here the integers. The new gadget brings with it some extra structure that can be exploited, after which it is eliminated and we fight our way back to the original problem.

Cyclotomic arithmetic

Several mathematicians noticed that it is possible to obtain a similar factorization of the expression $a^n + b^n$ by making use of the complex nth root of unity, usually denoted ζ. As in the previous section, if $a^n + b^n$ is an exact nth power, then this should imply that each factor is also a perfect nth power. In 1874 Lamé announced a proof of Fermat's Last Theorem based on this argument. Joseph Liouville, who had clearly been thinking about the matter for some time, promptly wrote to Lamé to point out that the proof only works if you can show that unique prime factorization holds for *cyclotomic integers*: polynomials in ζ with integer coefficients. Liouville's fears were confirmed when Ernst Kummer wrote to him to say that unique prime factorization fails for cyclotomic integers when $n = 23$. It looked as if this particular approach to Fermat's Last Theorem had hit an insuperable obstacle.

The ideal solution

One of the cardinal rules of mathematical research is not to give up on a good idea just because it doesn't work. Kummer, a theologian turned mathematician and a student of Gauss and Dirichlet, found a way to restore uniqueness of prime factorization, not just to the cyclotomic integers, but to *any* system of algebraic numbers. He got this idea when working on a completely different problem in number theory, known as the higher reciprocity laws. I mention only that these have to do with conditions under which the equation $a \equiv b^n \pmod{p}$ holds: the subject is extensive and important but would take us too far afield. He developed a theory of *ideal numbers*. In his original formulation, these are not numbers at all: they are the ghosts of non-existent numbers. An ideal number n is a relation between actual numbers that behaves like congruence modulo n, except that there is no n to be congruent modulo. Later it was found possible to exorcize the ghost in two different ways. An ideal number can be interpreted as a genuine number in a larger system of algebraic numbers, or as a *set* of numbers in the original system. The second interpretation found more favour, and led to a modern concept called an *ideal*. Kummer proved that although it is not always possible to factorize algebraic numbers into primes, it *is* possible to factorize them uniquely into prime ideal numbers. Armed with this new weapon (which must be wielded with the skill of a master duellist, parrying from ordinary to ideal numbers or riposting back again when the right opening appears),

Kummer proved Fermat's Last Theorem for exponents n that are 'regular' primes. There is a somewhat technical definition of what this means: suffice it to say that among primes less than 100 it covers all save 37, 59, and 67. Using additional arguments, Kummer and Dimitri Mirimanoff dealt with these cases too. It was conjectured that the result could be proved this way for infinitely many primes n: ironically, what is actually known is that the method *fails* for infinitely many primes! Of course, there may be other methods . . .

By extending these techniques, various mathematicians drove the frontiers further out; and by 1980 Wagstaff had shown that Fermat's Last Theorem is true for all n up to 125,000. Incidentally, this means that any counterexample, if it exists, must involve absolutely vast numbers, with at least a million digits—there's no hope of finding it by accident, or direct computation.

Euler's Conjecture

If it's not possible for two cubes to sum to a cube, might it be possible for *three*? It is; in fact $3^3 + 4^3 + 5^3 = 6^3$. Leonhard Euler conjectured that for all n it is possible for n nth powers to add to an nth power, but *impossible* for $n - 1$ to do so. However, Euler's conjecture is false. In 1966 L. J. Lander and T. R. Parkin found four fifth powers whose sum is again a fifth power:

$$27^5 + 84^5 + 110^5 + 133^5 = 144^5.$$

In 1988 Noam Elkies of Harvard University found the first counterexample to Euler's conjecture for *fourth* powers. Instead of looking for integer (whole number) solutions to Euler's equation $x^4 + y^4 + z^4 = w^4$ he divided out by w^4, which leads to a rational solution $r = x/w$, $s = y/w$, $z = t/w$ of the new equation $r^4 + s^4 + t^4 = 1$. (A rational number is a fraction p/q where p and q are integers.) Conversely, any rational solution of the new equation yields an integer solution of the original equation, by putting r, s, and t over a common denominator w. The new equation defines a surface in three-dimensional space, like a flattened ellipsoid. V. A. Demjanenko, found a rather complicated condition for a rational point (r, s, t) to lie on the related surface $r^4 + s^4 + t^2$

= 1, which involves t^2 rather than t^4. A solution to this will yield a solution to Euler's equation, provided t can be made a square. A series of simplifications shows that this can be done provided the equation

$$Y^2 = -31790X^4 + 36941X^3 - 56158X^2 + 28849X + 22030$$

has a rational solution. This equation belongs to a general class known as *elliptic curves*, widely studied by number theorists because they have very beautiful properties. The general theory of elliptic curves includes powerful tests for non-existence of solutions. They fail here, which hints—but does not of itself prove—that a solution may in fact exist. Encouraged by this indirect evidence, Elkies tried a computer search, and found one. Working backwards, he was led to a counterexample to Euler's conjecture for fourth powers, namely

$$2682440^4 + 15365639^4 + 187960^4 = 20615673^4.$$

The theory of elliptic curves also provides a general procedure for constructing new rational solutions from old ones, using the geometry of the curve. Using this, Elkies proved that rational points are dense on the surface $r^4 + s^4 + t^4 = 1$—that is, any patch of the surface, however tiny, must contain a rational point. There are, in short, *lots* of solutions. The numbers get very big, though. The second solution generated by this geometric construction is

$x =$ 1439965710648954492268506771833175267850201426615300442218292336336633;

$y =$ 4417264698994538496943597489754952845854672497179047898864124209346920;

$z =$ 9033964577482532388059482429398457291004947925005743028147465732645880;

$w =$ 9161781830035436847832452398267266038227002962257243662070370888722169.

After Elkies had discovered there *was* a solution, Roger Frye of the Thinking Machines Corporation did a computer search, and found the smallest possible one:

$$95800^4 + 217519^4 + 414560^4 = 422481^4.$$

New solutions for old

Let's take a closer look at the trick that Elkies used to find an infinite set of solutions, starting from just one. It's most easily explained for elliptic curves of the form

$$y^2 = ax^3 + bx^2 + cx + d.$$

Any straight line cuts such a curve at three points. If the coordinates of two are known, it is possible to calculate the coordinates of the third. If the first two have rational coordinates, the third does too. You can think of the third point as a sort of 'product' of the other two. Astoundingly, this product obeys reasonable algebraic laws, so the set of solutions forms what an algebraist would call an Abelian group. L. J. Mordell proved that there exist a *finite* number of rational solutions to the equation for any elliptic curve, such that every other rational solution can be obtained from them by repeated products of this kind. The group of solutions is 'finitely generated'. In particular, the equation has infinitely many rational solutions. Other Diophantine equations are much more miserly with their solutions, having only finitely many, or even none at all! For example there is an important theorem of Axel Thue and Carl Siegel, to the effect that a general class of equations of degree 3 or higher has only a finite number of solutions. A more specific example is W. Ljunggren's result of 1942 that there are exactly two solutions to the equation $x^2 + 1 = 2y^4$, namely $(x, y) = (1, 1)$ and $(239, 13)$. So the question is, how can we distinguish Scrooge from Santa Claus?

Equations with holes

The answer to this came from a very different direction: algebraic geometry. Algebraic geometers study (very generalized versions of) curves and surfaces defined by polynomial equations, called *algebraic varieties*. The main impetus for the theory came from nineteenth-century work on polynomials over the complex numbers. An equation $f(x, y) = 0$ in two real variables defines a *curve* in the (x, y)-plane. A similar equation in two complex variables defines an analogous object, a complex curve. However, the complex plane is two-dimensional whereas the real line is one-dimensional, so complex objects have a habit of doubling their dimensions in comparison with the corresponding real objects. It follows that a complex curve is two-dimensional over the reals, that is, it is really a *surface*. According to topologists, a general surface looks like a sphere with lots of handles glued to it, or

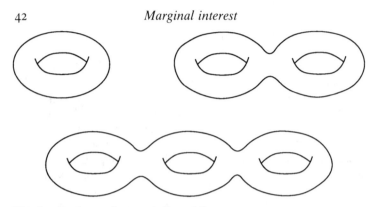

Fig. 2. Surfaces of genus 1, 2, and 3.

equivalently a torus (doughnut) with lots of holes. The number of holes is called the *genus* of the surface. So every complex equation has a number, its genus, associated with it. This topologically defined number may seem a rather arbitrary and obscure creature, but it is very important in algebraic geometry. It can, for example, be calculated arithmetically.

Mordell, in 1922, noticed that the only equations known to have infinitely many rational solutions are those of genus 0 or 1. For example the Pythagorean equation has genus 0; elliptic curves genus 1. Every equation treated by Diophantus has genus 0 or 1! Here at last was a semblance of order and generality. Mordell conjectured that every equation of genus 2 or more has only finitely many rational solutions, possibly none. The Fermat equation has genus $(n - 1)(n - 2)/2$, which is greater than 1 when n is greater than 3. So *one single special case* of the Mordell Conjecture implies that for any $n > 3$, there are only finitely many solutions (if any) to the Fermat equation. This shows what a powerful result the Mordell Conjecture would be. The next step, after proving finiteness of the set of solutions, would be to put a definite bound on their size—'make the theorem effective'. After that, in principle routine computation would show whether any solutions at all exist, and if so, find them. Of course this is all pie in the sky, but so was the Mordell Conjecture.

The evidence in favour of the Mordell Conjecture was very slight; so slight that André Weil, one of the world's leading number theorists, remarked that 'one cannot perceive a single good reason to bet for or against it'. Weil is a frequent critic of the overuse of the word 'conjecture' to dignify wild guesses that mathematicians *hope* might turn out to be true. Ironically, it was some conjectures due to Weil that opened up a path to a proof of the Mordell Conjecture.

The fruits of boredom

Weil tells us that in 1947, while in Chicago,

I felt bored and depressed, and, not knowing what to do, I started reading Gauss's two memoirs on biquadratic residues, which I had never read before. The Gaussian integers occur in the second paper. The first one deals essentially with the number of solutions of equations $ax^4 - by^4 = 1$ in the prime field modulo p, and with the connection between these and certain Gaussian sums; actually the method is exactly the same that is applied in the last section of the *Disquisitiones* to the Gaussian sums for order 3 and the equations $ax^3 - by^3 = 1$. Then I noticed that similar principles can be applied to all equations of the form $ax^m + by^n + cz^r \ldots = 0$, and that this implies the truth of the so-called 'Riemann Hypothesis' for all curves $ax^n + by^n + cz^n = 0$ over finite fields. This led me to conjecture about varieties over finite fields. . . .

The Weil Conjectures can be described as basic contributions to the following question: given a prime p, how many solutions does a given Diophantine equation have modulo p? This is important for ordinary integer solutions of Diophantine equations, because any solution in integers can be 'reduced' to give a solution modulo p. So, for example, if there are no solutions modulo p, there are no solutions at all! In practice there are problems in 'gluing together' the information coming from reduction modulo different primes; and Weil developed a very beautiful idea. It is known that for any prime power p^n there exists a *finite field* having exactly p^n elements. A field is a system with a self-consistent arithmetic permitting division. Given an equation, let a_n be the number of solutions that it has in this field. There is a gadget called the *zeta-function* of the equation which captures, in a single object, all the numbers a_n. Define an *algebraic variety* to be the set of solutions of not just one equation, but a whole system. Then Weil conjectured, on the basis of his investigations on curves, that for a general algebraic variety V, three things hold:

(1) The zeta-function of V can be computed in terms of a special finite set of algebraic integers.

(2) There is a 'functional equation' relating the zeta-functions of s and s/p^n.

(3) (The 'Riemann Hypothesis' for V) The algebraic integers in (1) all lie on a circle in the complex plane, whose centre is the origin, and whose radius is $p^{n/2}$.

All this may seem pretty technical, but the point is that you get very detailed information about the number of solutions to a system of Diophantine equations over a finite field.

The topological connection

How did Weil come to these conjectures? They weren't just guesswork; he had a strong suspicion that they should hold. They 'smelt right'. The reason was an analogy with topology. The idea is to apply a result called the *Lefschetz trace formula*, which counts the number of fixed points of a topological transformation. The transformation here is the *Frobenius map*, which changes x to x^p. The upshot of all this is Weil Conjecture (1). Conjecture (2) comes from a topological theory of 'duality'. Number (3) is much deeper, involving other conjectures due to Solomon Lefschetz and William Hodge. There's one snag: the Lefschetz trace formula only works for topological spaces, and finite fields have no sensible topological structure. So Weil's analogy can't be made rigorous. In fact it would have been dismissed out of hand, except that *it gives the right result for curves*. Eventually the Weil Conjectures were proved in 1975 by Pierre Deligne, using rather different (and difficult) methods. They have since proved to be absolutely fundamental to many problems involving algebraic geometry.

A surfeit of conjectures

Ever since the nineteenth century algebraic geometers have exploited an analogy between algebraic number fields and 'function fields', whose elements satisfy an equation whose coefficients are not rational numbers, but rational functions

$p(x)/q(x)$ where p and q are polynomials. The first evidence that Mordell might be right came in 1963 when Yuri Manin proved the analogous result for function fields. H. Grauert obtained the same result independently in 1965. Meanwhile in 1962 Igor Shafarevich had made a conjecture about 'good reduction' for curves—the process whereby an equation over the integers is reduced modulo a prime. (Instead of solving, say, $x^3 = 22$, one looks at equations such as $x^3 \equiv 22 \pmod 7$. This has the solution $x \equiv 2 \pmod 7$ and tells us something about the original x. By using many primes in place of the 7, further information can be gained.) His idea was that there should be only finitely many different types of curve, of given genus, having good reduction for all but a fixed finite set of primes. The important feature is the finiteness of the set of curves. And the pace hotted up in 1968 when A. N. Parshin proved that the Shafarevich Conjecture implies the Mordell Conjecture. As you may by now have noticed, algebraic geometry is rife with conjectures. In 1966 John Tate stated yet another, about 'abelian varieties', a generalization of elliptic curves to arbitrary dimensions. By the early 1980s experts in Diophantine equations were beginning to suspect that Mordell was right, that a proof was lurking not far away, and that Tate's conjecture was also involved. And in 1983 Gerd Faltings found a way to prove not only Mordell's Conjecture, but also those of Shafarevich and Tate. His method can be viewed as a very elaborate variation on Fermat's 'infinite descent'. Faltings first proves the Tate Conjecture. Combining that with Deligne's results on the Weil Conjectures, he establishes the Shafarevich Conjecture. And now Parshin's work on the Mordell Conjecture completes the proof. (A different proof was found very recently by P. Vojta.)

Regarding Fermat's Last Theorem, a gap remains. Finitely many solutions is not the same as none. For all we can tell from Faltings's Theorem, the Fermat equation might have millions of solutions for any n! However, D. R. Heath-Brown has recently proved that the proportion of integers n for which it has no solutions approaches 100 per cent as n becomes large, so Fermat's Last Theorem is 'almost always' true. Can we drop the 'almost'? That's for the next generation of mathematicians to decide.

Easy as abc

A totally new approach to the Fermat Conjecture—and much
else—was initiated by R. C. Mason in 1984. He discovered a
new property of polynomials, with amazing consequences. Its
generalization to whole numbers—if true—would have even
more amazing consequences. Recall that a *zero* of a polynomial $p(x)$ is a solution of the equation $p(x) = 0$. Mason's
theorem is this. Suppose we have three polynomials $a(x)$, $b(x)$,
$c(x)$, without common factors, such that $a(x) + b(x) = c(x)$.
Then the number of distinct zeros of the three polynomials is
at least one greater than their largest degree.

This may seem a highly technical theorem, but here's an
immediate consequence: Fermat's Last Theorem is true for
polynomials. For suppose $a(x)^n + b(x)^n = c(x)^n$ for all x. By
Mason's Theorem, the degree of $a(x)^n$ is less than or equal to
the sum of the degrees of a, b, and c, minus one. The same
goes for $b(x)^n$ and $c(x)^n$. Adding, we get n times the sum of
the degrees being less than or equal to three times their sum,
minus three. This implies that n is at most two.

Inspired by this result, Maser and Oesterle formulated an
analogue for integers. It's a little more complicated. Suppose
that a, b, c, are relatively prime integers such that $a + b = c$.
Let ε by any positive number. Then there exists a number C,
depending on ε, such that a, b, and c are all smaller in
absolute value than Cr^ε, where r is the product of the distinct
prime factors of a, b, and c. This is known as the *abc conjecture*. It remains unproved, but its consequences are spectacular.

One is that Fermat's Last Theorem is true for *all* sufficiently
large exponents n. That is, there exists some value N such that
when $n > N$ there are no non-trivial solutions to Fermat's
equation. A second is that there are infinitely many primes p
such that $2^{p-1} \not\equiv 1 \pmod{p^2}$, a classically conjectured gloss on
Fermat's Little Theorem that has never been proved. A third
is that for large k the Diophantine equation $x^2 = y^3 + k$ has
fewer than $k^{2+\varepsilon}$ solutions for any positive ε. It also implies
two more technical conjectures due to Marshall Hall and P.
Vojta. The second of these is related to Faltings's work. Of
course, nobody knows how to prove the *abc* conjecture. But it
certainly puts the entire area in a new light.

4

The neglected book of Euclid

The discovery of incommensurable ratios is
attributed to Hippasus of Metapontum. The
Pythagoreans were supposed to have been at sea
at the time and to have thrown Hippasus over-
board for having produced an element in the
universe which denied the Pythagorean doctrine
that all phenomena in the universe can be
reduced to whole numbers or their ratios.

Morris Kline

One of the most famous books in the whole of mathematics's
chequered history is Euclid's *Elements*, a series of texts on
geometry dating from about 300 BC. Euclid set a standard
in logical argument that was not surpassed for 2,000 years.
Although to modern eyes his logic has gaps, nevertheless his
axiomatic approach provides a model for much of today's
mathematics. Large parts of the *Elements* were used in school-
teaching, virtually unchanged, from the Middle Ages to the
end of the nineteenth century. The topics taught included
congruent triangles, Pythagoras's theorem, the construction of
polygons, and the classification of the five regular solids. But
one of Euclid's books was hardly ever taught: the tenth. It's
much more obscure than the other books: on the face of it, a
jumbled compilation of technical results with no organized
theme. There were probably few teachers who could see any
point to Book Ten, let alone appreciate its role in the scheme
of mathematics. Its 'feel' is quite unlike that of the remaining
twelve books. And the reason is that it isn't really about
geometry at all, but arithmetic. It is evidence of a lengthy
struggle with some of the deeper properties of the real number
system.

The story is an intriguing one, because it brings together
several ideas that in various guises have moulded the devel-

opment of mathematics. In particular, it shows how possession of a well-developed body of technique can side-track subsequent research, while obscuring the existence of a simpler and clearer point of view. In other words, when you can do something successfully, in no matter how cack-handed a fashion, it's a lot easier to carry on doing it, rather than develop a better understanding of what's really going on. Indeed, practitioners of a complicated technique are likely to revel in its complications, once they have mastered them. The same occurs fairly often in modern research. It's unfortunate, because the essence of good mathematics is to penetrate to the heart of a problem: a solution alone is not the ultimate goal.

Sacrifice to the irrational

Although mathematics is not solely, nor even primarily, about numbers, these play a fundamental role. Mathematicians distinguish many types of number. The *natural numbers* are the traditional positive whole numbers 0, 1, 2, 3, . . . and so on. The *integers* are positive or negative whole numbers. The *rational numbers* (or just 'the rationals') are ratios of whole numbers—that is, fractions. The *real numbers* ('the reals') are all numbers representable by a decimal expansion, terminating or non-terminating. The *complex numbers* arise from the reals when we allow -1 to have a square root.

Historically, each new extension of the number system has been achieved only after lengthy philosophical battles. Invariably the battles were won, not by the excellence of the intellectual argument in favour of extending the system, but by the overwhelming utility of the results obtained by so doing. Numbers are so closely allied to certain aspects of the natural world that we tend to think of them as something unique and almost physical. It is only when they are analysed more deeply that it becomes clear that they are an invention of the human mind—a method whereby our brains can *model* aspects of Nature. They are not Nature herself.

By the time of the Pythagoreans, the number system had developed to what we would now describe as the stage of rational numbers: whole numbers and their ratios. Whole numbers are easy to visualize and manipulate: all you need is a large supply of pebbles to use as counters. Fractions are also reasonably approachable: to obtain the number $\frac{2}{3}$ you take an object, such as a bagful of sand, divide it into three equal heaps, and select two of them. There are infinitely many fractions, and they allow an infinitely fine division of the series of numbers. For purposes such as trade, land-measurement, and astronomy, the rationals are more than sufficient. Even today, the most accurate scientific measurements can be expressed to no more than about ten decimal places; so rationals with denominators up to 10 billion or so suffice to describe the results of experiment. They do not, however, measure up to the requirements of *theory*.

Investigating the logical basis of known geometry, the Pythagoreans made an astounding discovery. In geometry, numbers naturally arise as the lengths of straight lines. Moreover, the Greeks knew that the arithmetical operations can be effected by geometrical constructions. For example, two numbers may be added by placing the corresponding lengths end to end. It seemed obvious that to each length there corresponds a number (meaning rational number), and to each (rational) number a length. With the rationals being so finely spaced, there can surely be no gaps—no lines that fail to have a rational length. However, the Pythagoreans managed to prove that certain easily constructible lines have lengths that correspond to *no rational number whatsoever*. They called them *irrational*. For example, the diagonal of a unit square (whose length we would now say is $\sqrt{2}$) is such a line. The *golden number* $\tau = (1 + \sqrt{5})/2$, which occurs, for example, in the geometry of pentagons, is another. There seem to be three choices: deny that this diagonal exists (which leaves geometry in ruins); accept that lines may not have lengths (ditto); or accept that lengths may not correspond to numbers (which ruins arithmetic). Worse, any theorems whose proofs have tacitly assumed lengths to be rational are fallacious. For example, the theorem that the ratio of the areas of two squares is the same as the square of the ratio of their sides.

Ratio vincit omnia

The Greek solution was to throw out arithmetic and take the lines themselves as being synonymous with their lengths. Then areas are synonymous with the corresponding squares, and so on. A *ratio* then becomes some sort of relation between pairs of lines. The main reason this trick works is that all of the really basic theorems are about ratios. You don't need to know what a ratio *is*, you just need to know when two of them are equal, and if not, which one is bigger. The Greek answer to this, supplied by Eudoxus, is at the start of Book Five of the *Elements*. A modern version of his definition might be this: 'to distinguish two irrationals you must find a rational between them.' The original is far more unwieldy and works not with numbers but with ratios. To the modern eye it immediately suggests the idea that irrationals are to be treated in terms of approximating rationals. In the hands of the Greeks it led to much the same idea, but in a more cumbersome form.

Exhaustion

Suppose you have two triangles that are the same shape, but one is twice the size of the other (that is, its sides are twice the length). Then we know that its *area* is four times as big. Areas vary with the square of the size. The Greeks could prove this for triangles; and by gluing a lot of triangular pieces together

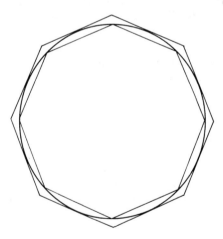

Fig. 3. The ancient Greek method of exhaustion approximates a circle by inner and outer polygons whose areas differ from each other by an arbitrarily small amount. In this way the area of the circle can be found.

they could prove it for polygons. But even as simple and basic a figure as a circle can't be made up out of triangles. Nevertheless they managed to extend the proof to circles and other figures by *approximating* them with polygons. There is an inside polygon and an outside polygon for the first circle, and a similar pair for the second circle. The inside polygons have areas proportional to the squares of their sides, and the same goes for the outer ones. The difference between the inner and outer polygon can be made as small as we please, and the circle itself is sandwiched in between.

In this situation the Eudoxus approach applies. Either the ratio of the areas of the circles is equal to the ratio of the squares of their sizes, or it's bigger, or it's smaller. If smaller, then it must be smaller by a definite amount; this leaves room to squeeze an approximating polygon into the gap, showing that it also goes wrong for the polygon. But that's absurd, because the result is known to be true for a polygon. The same argument applies if it's bigger. Conclusion: only equality is possible! The full-blooded Greek version of this method is called *exhaustion*, because the approximating polygons 'exhaust' the entire area. At first sight the method appears very clumsy, but despite its clumsiness, its quite easy to use after a bit of practice. And that's what trapped the Greeks into a rigid mould of thinking. It prevented them from devising a decent form of arithmetic or algebra; it even distorted the range of curves and surfaces that their geometry tried to handle. What is remarkable is how much good work they did under this self-imposed handicap. (To be fair to the Greeks, it would have been very hard to develop a full understanding of the subtleties of the real numbers without the geometric insights that their work provided.)

The legacy of Book Ten

What the obscure Book Ten seems to be about is an attempt to classify different types of irrationals. Basically, the only irrationals that you can get to grips with geometrically are those that involve square roots. So there are simple ones, like $\sqrt{7}$, and more complicated ones like:

$$\sqrt{[3\sqrt{(7\sqrt{31} - 19\sqrt{2})} + 11\sqrt{(7\sqrt{31} + 19\sqrt{2})}]}.$$

Book Ten tries to grapple with these complexities. Its results aren't very important today, but the study of different *kinds* of irrationality certainly is. And one important tool, *continued fraction expansion*, can be traced right back to Book Ten.

A typical continued fraction looks like this:

$$1 + \cfrac{1}{3 + \cfrac{1}{2 + \cfrac{1}{4}}}$$

It's not as daunting as it looks provided you work it out in the right order. Starting at the bottom end, we have 2 + 1/4 = 9/4. The reciprocal of this is 4/9, and adding 3 we get 31/9. Take the reciprocal again, getting 9/31, and add the initial 1 to get 40/31. That's the answer. As a matter of notational convenience the sprawling formula can be abbreviated to [1; 3, 2, 4], and both the printer and my word-processor will be much happier if we all agree to this.

Given any number, you can find its continued fraction. For a rational number such as 40/31, the fraction eventually stops. For an irrational it does not, and this provides a clear distinction between the two types of number, though not one that's always easy to use! The expansions of some important irrationals look like this:

$$\pi = [3; 7, 15, 1, 292, 1, 1, 6, 1, 3, \ldots]$$
$$e = [2; 1, 2, 1, 1, 4, 1, 1, 6, 1, 1, 8, \ldots]$$
$$\sqrt{2} = [1; 2, 2, 2, 2, 2, 2, 2, 2, 2, 2, \ldots]$$
$$\sqrt{3} = [1; 1, 2, 1, 2, 1, 2, 1, 2, 1, 2, 1, \ldots]$$
$$\tau = [1; 1, 1, 1, 1, 1, 1, 1, 1, 1, 1, 1, 1, \ldots].$$

Here e is the base of natural logarithms, and τ is the golden number $(1 + \sqrt{5})/2$ mentioned earlier.

Some patterns

Let's take a closer look at π. If we chop off the continued fraction very early we get [3; 7] which evaluates as 22/7, a fraction that every schoolchild knows. If we chop it off a couple of steps later we get [3; 7, 15, 1] = 355/113, another famous approximation to π. Numerically we have

$$22/7 = 3.1428571\ldots$$
$$355/113 = 3.1415929\ldots$$
$$\pi = 3.14159265\ldots$$

and we notice that retaining more terms in the expansion gives closer results. It can be proved that the continued fraction expansion provides the *best* rational approximations possible.

There are some fascinating patterns—or absences of pattern—in the results listed. The most obvious case is τ, where all the terms are 1! There are also repeating sequences in $\sqrt{2}$ and $\sqrt{3}$. All three of these involve nothing worse than square roots of rationals; and it was established in the eighteenth century that this is no coincidence. Repeating continued fraction expansions correspond to 'quadratic irrationals' $a + \sqrt{b}$ where a, b are rational.

You may also notice that e shows a pattern, but not a repeating one. The pattern requires the fraction to continue indefinitely, and it follows that e must be irrational. Since the continued fraction is not periodic, it cannot be a quadratic irrational either. Euler observed this in 1737 as part of a general investigation of continued fractions. Johann Lambert used continued fractions to prove that if x is rational then e^x and tan x. are not. It follows that π is irrational. Mathematicians would dearly like to know what patterns in continued fraction expansions correspond to what numbers. For example, can we say anything about *cubic* irrationalities (such as $\sqrt[3]{2}$)? For what numbers is there an upper limit to the size of the terms in the continued fraction? These questions are important in all sorts of places, notably the theory of dynamical systems. Very little is known.

Some are more irrational than others...

The continued fraction approach has led us to distinguish the quadratic irrationals, with their periodic patterns, but there are other ways to classify irrationals. As mentioned in chapter 3, one of the most basic is to divide them into *algebraic* numbers, which satisfy a polynomial equation with rational coefficients, and *transcendental* numbers, which don't. The numbers $\sqrt{2}$, $\sqrt{3}$, τ, and $\sqrt[3]{2}$ are algebraic. A polynomial equation satisfied by π or e hardly springs readily to mind, suggesting that they are transcendental. On the other hand, there are unexpected formulae in mathematics, such as Euler's remarkable

$$e^{\pi\sqrt{-1}} + 1 = 0,$$

so it's unwise to jump to conclusions. As it turned out, e and π *are* transcendental, but the proof took a while to arrive. In fact, until 1844 nobody knew whether any transcendental numbers existed at all. In that year, Liouville proved a theorem to the effect that algebraic irrationals cannot be approximated very well by rationals. It's then an easy task to find an irrational with unusually good rational approximations, which perforce must be transcendental. An example is

$$1.101001000010000000010000000000000000001 \ldots$$

where the number of 0s between consecutive 1s doubles at each stage. This is still a far cry from proving transcendence of any 'naturally occurring' number. That step was taken by Charles Hermite in 1873, who proved that e is transcendental. He wrote: 'I do not dare to attempt to show the transcendence of π. If others undertake it, no one will be happier than I about their success, but believe me, my dear friend, this cannot fail to cost them some effort.' Ferdinand Lindemann succeeded in this task in 1882 by a method very similar to Hermite's, making use of Euler's formula mentioned above.

Squaring the circle

Why were people so keen to prove π transcendental? The answer is that it knocks on the head a very notorious problem. It is normally grouped together with two other problems: three left-overs from Greek geometry, which ask for constructions, using only an unmarked ruler and a pair of compasses, for

(a) A square whose area is the same as that of a given circle;

(b) An angle one third the size of a given angle;

(c) The side of a cube that is twice the volume of a given cube.

They are known, respectively, as the problems of squaring the circle, trisecting the angle, and duplicating the cube. The transcendence of π is important because it proves that it is impossible to square the circle. The other two problems are also insoluble under the stated restrictions. Before I explain why, it is worth examining the history of these problems in a little more detail, because the usual version of the story tends to produce serious misconceptions.

The first point to make is that *the ancient Greeks could solve all three problems*. They knew how to find π, they knew about $^3\sqrt{2}$, and they encountered no difficulty, either theoretical or practical, in splitting an angle into three equal bits. Not if they respected the restriction to an unmarked ruler and a pair of compasses, of course; but those restrictions were not widely adhered to by the great mathematicians of ancient Greece. Eutocius, a commentator from the sixth century AD, describes a dozen different methods for duplicating a cube. Several are based on so-called neusis-constructions, which involve sliding a *marked* ruler—a ruler with a distinguished point, or several, on its edge—along some configuration of lines until some particular condition holds. Others make use of conic sections; and there is a stunning three-dimensional construction that makes use of a cylinder, a cone, and a torus.

There are fewer reports of methods for trisecting angles, possibly because there exists an extremely simple and obvious

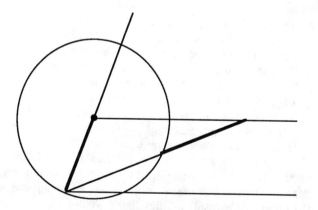

Fig. 4.   Trisecting an angle with a marked ruler. Lay the ruler so that the bold lengths are equal: then the ruler trisects the angle at lower left.

neusis-construction. A general method for dividing angles by any whole number whatsoever, attributed to Hippias, makes use of a transcendental curve called the quadratrix.

The quadratrix can also be used to square the circle; and Archimedes, in a surviving fragment that exists only in a much later edition, proves a result which in current terminology states that π lies between $3\frac{10}{71}$ and $3\frac{1}{7}$. David Fowler argues that Archimedes may have been trying to evaluate the continued fraction of π, and that the Greeks used continued fractions fairly systematically as a way of forming hypotheses on the rationality or irrationality of particular numbers. If so, Archimedes had got as far as [3; 7, something $\geqslant 10$]. Had he got as far as [3; 7, 15, 1, something $\geqslant 200$] he might well have begun to wonder whether π might actually be rational, for such unusually large terms normally appear in incomplete developments of rationals.

At any rate, the ancient Greek master mathematicians were perfectly happy using marked rulers and transcendental curves if they needed them. Where, then, did the restriction to constructions with unmarked ruler and compasses come from? It appears to be Eutocius, a mere hack, who explicitly and not very politely criticized the great mathematicians of earlier times for not respecting restrictions that he imposed several centuries later. In short, the now notorious problems 'bequeathed to us by the ancient Greeks' were not actually problems that the *ancient* Greeks, the greats, the *real* mathematicians, ever worried about. It is not unusual for the history of mathematics to be rewritten in this manner.

Be that as it may, later mathematicians took Eutocius's restrictions seriously, and wondered whether constructions might exist that obeyed them. Eventually they proved that there are none, by invoking algebraic methods. Any geometric construction that obeys Eutocius's restrictions can be broken down into elementary steps, and intepreted as a series of solutions to linear or quadratic equations. Therefore any length that can be constructed in the prescribed manner must solve a polynomial equation—indeed, one of a fairly special kind. Lindemann's result implies that π is not such a length; therefore it is impossible to square the circle using unmarked ruler and compasses.

The other two problems go the same way. Duplicating the cube amounts to solving the cubic equation $x^3 = 2$, and an argument rather simpler than Lindemann's shows that this cannot be reduced to a series of quadratics. Trisecting an angle is also equivalent to solving a cubic equation: for an 'arbitrary' angle, this again cannot be reduced to a succession of quadratics.

Here we see mathematics playing one of its typical tricks. The original problem—to square the circle—is obviously one in geometry. You might expect its solution also to be geometric. Translating it into algebraic terms, via coordinate geometry, leads to a reformulation whose solution you might expect to be algebraic. But Lindemann's proof is analytic; it uses methods of calculus, in particular Euler's formula connecting e and π (whose proof uses infinite series). The long-sought resolution of the problem does not—indeed cannot—arise from a direct geometrical attack. It is a useful taxonomic device to divide mathematics up into labelled specialities, but anyone who imagines that the subject itself respects those divisions is ignoring the lessons of history.

A proof that Euler missed

In 1737 Euler noticed that he could use the series

$$\zeta(x) = 1 + 2^{-x} + 3^{-x} + \ldots$$

to prove results about prime numbers—for example that the sum of their reciprocals is infinite, showing not just that there exist infinitely many primes, but that they are not too thin on the ground. (In 1859 Bernhard Riemann embarked on a deep extension of Euler's work, founding the analytic theory of prime numbers. The function $\zeta(x)$ is now known as the Riemann zeta function). After enormous trouble Euler managed to sum the series for certain values of x. In 1734 he found that $\zeta(2) = \pi^2/6$. Later he proved that for all *even n*, $\zeta(n)$ is a rational multiple of π^n. We may deduce from this that $\zeta(n)$ is irrational (indeed transcendental) for all even n. Until very recently, nobody could say anything along these lines for odd n, and the whole problem looked pretty hopeless. You may imagine the reaction when, at the Journées Arithmétiques de Marseille-Luminy, in June 1978, R. Apéry of the University of Caen was scheduled to talk 'On the irrationality of $\zeta(3)$'. Alf van der Poorten, who was there, reported the talk as follows: 'Scepticism was general. The lecture tended to strengthen this view to rank disbelief. Those who listened casually, or were afflicted with being non-Francophone, appeared to hear only a sequence of unlikely assertions.' Van der Poorten then lists some bizarre formulae leading up to a sequence of rational approximations to $\zeta(3)$ that converges to it so rapidly that (by Liouville's result) it cannot be rational. He continues:

I heard with some incredulity that, for one, Henri Cohen believed these claims might well be valid. Very much intrigued, I joined Hendrik Lenstra and Cohen in an evening's discussion in which Cohen explained and demonstrated most of the details of the proof. We came away convinced that Professeur Apéry had indeed found a miraculous and magnificent demonstration of the irrationality of $\zeta(3)$. But we remained unable to prove a critical step.

The problem involved a sequence constructed by Apéry during the course of this proof, which he claimed consisted of integers. There seemed to be no reason why this should be true. Indeed at first sight it was 'obviously' false, because the formulae Apéry used were so weird. However... 'In Marseille, our amazement was total when our HP-67 calculators kept on producing integer values.' But only Apéry, it seemed, knew why; and he wasn't telling. Then, two months later, at the Helsinki meeting of the International Congress of Mathematicians, Cohen and van der Poorten mentioned their difficulty to Zagier. 'With irritating speed' he filled the gap in their argument. This established that Apéry was right and so were his bizarre formulae.

Van der Poorten sums the episode up:

What on Earth is going on here? Apéry's incredible proof appears to be a mixture of miracles and mysteries. Here we have, apparently, the tip of an iceberg. Does the complete berg look like this? For my part I incline to the view that much of what has been presented constitutes a mystification rather than an explanation. Most startling of all though should be the fact that Apéry's proof has no aspect that would not have been accessible to a mathematician of 200 years ago.

If only, of course, that mathematician had possessed Apéry's insight.

5

Parallel thinking

The study of 'non-Euclidean' geometry brings nothing to students but fatigue, vanity, arrogance, and imbecility. 'Non-Euclidean' space is the false invention of *demons*, who gladly furnish the dark understandings of the 'non-Euclideans' with false knowledge. The 'non-Euclideans', like the ancient sophists, seem unaware that their understandings have become obscured by the promptings of the *evil spirits*.

Matthew Ryan

Sydney Smith tells of two disputatious women with a garden fence between them. 'Those women will never agree', said Smith; 'they are arguing from different premises.' This reflects the lesson, painfully learned by philosophers and logicians alike, that you can't get something for nothing. Any chain of logical argument has to start with unproved hypotheses. As Aristotle said, 'It is not everything that can be proved, otherwise the chain of proof would be endless. You must begin somewhere, and you start with things admitted but undemonstrable.' Euclid was aware of this, and when he wrote the *Elements* he started out by listing his assumptions. Most of them are unobjectionable enough: things like 'objects equal to the same thing are equal to each other.' But among them is one whose intuitive appeal is much more debatable: 'If a straight line falling on two straight lines makes the interior angles on the same side less than two right angles, then the two straight lines, if produced indefinitely, meet on that side on which are the angles less than two right angles.' With a little work it can be seen that this assumption is equivalent to a more readily intelligible one: given a line, and a point not on the line, there is a unique parallel to the line through the given point. A *parallel*, of course, is a line that does not meet the original line however far it is produced, that is, extended.

Commentators on Euclid generally regarded the need to make such an assumption as a blemish. Jean Le Rond d'Alembert called it a scandal. Proclus, in about AD 450, criticized it roundly: 'This ought to be struck out altogether; for it is a theorem involving many difficulties, which Ptolemy, in a certain book, set himself to solve. The statement that since the two lines converge more and more as they are produced, they will sometime meet, is plausible but not necessary.' It is clear that the Greeks had thought long and hard about the nature of the assumption. Innumerable attempts were subsequently made to prove that it follows from the rest of Euclid's assumptions: none was successful. Euclid's judgement was vindicated when it began to be clear that there exist consistent versions of geometry in which his *parallel axiom* does not hold. Instead of a single Geometry there is a variety of geometries, in which different theorems are true. Their discovery took a long time, because everybody assumed that geometry somehow exists as an object of Nature, not just as a construct of the human mind. Eventually it was realized that, while it's perfectly sensible to ask 'what is the true geometry of Nature?', it's not a question that can be answered by mathematical reasoning alone.

Euclid without flaws

The earliest work on the problem has as its goal a *proof* of the property of parallels asserted by Euclid. A small number of mathematical problems seem to attract crackpots: squaring the circle, trisecting the angle, Fermat's Last Theorem, and the parallel axiom. As well as the lunatic fringe, there was some excellent work by people who *thought* they had found a proof, but who in reality had replaced Euclid's assumption by some logically equivalent, unproved one. These assumptions include:

Three points either lie on a line or on a circle.
The area of a triangle can be as large as we please.
The angles of a triangle add up to 180°.
Similar figures exist.

Others recognized that they had only found a logical replacement for Euclid's axiom, but felt their version was more intuitive.

The first serious attack was made by Girolamo Saccheri, who published a two-book work, *Euclid freed of every Flaw*, in 1733. Saccheri attempted a proof by contradiction, based on the following construction. Draw a straight line, erect two perpendiculars of equal length, and join them by a new line. Consider the angles (they are equal) made by this line with the perpendiculars. There are three cases:

(a) They are *right* angles.
(b) They are *obtuse* (greater than a right angle).
(c) They are *acute* (less than a right angle).

If you like, Saccheri's idea is that we try to build a rectangle and then ask whether we have succeeded. He showed that (a) is the same as the parallel axiom. The idea was to show that each of (b) and (c) leads to an absurdity, hence is false. Saccheri proceeded to make a series of deductions. Book 1 demolishes (to its author's satisfaction) the obtuse angle, and ends with the defiant statement that 'the hypothesis of the acute angle is absolutely false, being repugnant to the nature of the straight line'. Saccheri was whistling in the dark. That he wasn't convinced himself is suggested by the existence of Book 2, which attempts to shore up the argument. A crucial proposition, number 37, is given three proofs: all are fallacious. One involves infinitesimals; one, if correct, would prove that all circles have the same circumference; and one invokes physical ideas of time and motion. After this futile wriggling on his own hook. Saccheri comes to Proposition 38: 'The hypothesis of the acute angle is absolutely false because it destroys itself.' He then sums everything up by describing the refutation of the hypothesis of the obtuse angle as being 'as clear as midday light', but notes that the falsity of the acute angle depends on that worrisome Proposition 37.

Fifty years later Lambert tried an approach very similar to Saccheri's, proving a large number of theorems which, although very strange to Euclidean eyes, do not give logical contradictions. Among his results on the hypothesis of the acute angle there is a formula relating the area of a triangle to the sum of its angles:

$$k^2(\pi - (A + B + C))$$

where k is a certain constant. This is very similar to a theorem in spherical geometry, where the area of a triangle on a sphere of radius r is

$$r^2((A + B + C) - \pi).$$

Replacing r by $k\sqrt{-1}$ Lambert hazarded the guess that the corresponding geometry is that of a sphere of imaginary radius. What he meant by this—if anything—is unclear; but we can detect in his work the germ of the idea that the parallel axiom may fail for the geometry of a surface.

Parallel discoveries

All of this early work was motivated by a desire to demolish any alternatives to Euclid. But by the early nineteenth century a bolder viewpoint emerged: to devise self-consistent geometries that differ from Euclid's, in particular with regard to parallels. The main ideas seem to have been discovered independently by three people: János Bolyai, Nikolai Ivanovich Lobachevskii, and Gauss.

Bolyai was a dashing young artillery officer who reputedly fought, and won, thirteen consecutive duels in one day. Lobachevskii was a lecturer at Kazan University. Gauss was the leading mathematician of his (perhaps any) age. Bolyai's father Farkas was a friend of Gauss's, and had worked on the parallel axiom himself. He advised his son not to waste his time on such a poisonous problem. But in 1823 the son wrote to his father of his 'wonderful discoveries', saying 'I have created a new universe from nothing.' Meanwhile Lobachevskii was busy: he lectured on parallels in 1826, and published a lengthy paper in Russian in 1829 which established, for the first time in print, an alternative to Euclid. It bore the unhappy name *Imaginary Geometry*. Bolyai published his work in 1832.

Gauss, in one of the more curious passages of his life, then wrote to the elder Bolyai to the effect that he was unable to praise the son's work because he, Gauss, had obtained identical results much earlier (but not put them into print). Gauss's refusal condemned Bolyai to a life of mathematical obscurity, and the merits of his work were not recognized until after his death. Lobachevskii fared little better: he died blind and in poverty. There is evidence to back up Gauss's claim: in 1824 he wrote to Franz Taurinus saying that the hypothesis of the acute angle 'leads to a peculiar geometry completely different from ours, a competely self-consistent geometry that I have developed for myself perfectly satisfactorily, so that I can solve any problem in it . . . If at some time I have more leisure than now, I may publish my investigations.' In fact he never did, apparently worried that the public wouldn't understand: 'I fear the cries of the Boeotians . . .'

Geometries within geometries . . .

All three workers had convinced themselves that they had found an alternative, self-consistent type of geometry. They felt that way because everything seemed to fit together sensibly. But a complete proof that the system will *never* lead to a contradiction was lacking. It is doubtful if this lack was really noticed; certainly nobody worried too much about it. The final step was taken by Eugenio Beltrami in 1868. He was investigating the geometry of surfaces. On a curved surface, the analogue of a line is a *geodesic*—the shortest curve between two points. Beltrami introduced a surface called the *pseudosphere*, and showed that the geometry of its geodesics satisfies Saccheri's hypothesis of the acute angle. It also satisfies the rest of Euclid's axioms. It is Bolyai/Gauss/Lobachevskii's non-Euclidean geometry in disguise. It follows that non-Euclidean geometry can never lead to a contradiction—because, if it does, the pseudosphere cannot exist. And the pseudosphere is a construction in *Euclidean* geometry! More accurately, what we now know is that the only way non-Euclidean geometry can contradict itself is for Euclidean geometry to contradict itself. If Saccheri's programme had succeeded, it would have demolished Euclidean geometry, not perfected it!

Beltrami had found what we now call a *model* for non-Euclidean geometry: a mathematical system whose objects, suitably interpreted, obey all of Euclid's axioms except that of parallels. The existence of such a model rules out any possibility of proving the parallel axiom from the others. For such a proof, interpreted within the model, will force the model to obey its interpretation of the parallel axiom as well. As David Hilbert said one day in the Berlin railway station: 'One must be able to say at all times—instead of points, straight lines, and planes—tables, chairs, and beer mugs.' The idea behind this is that only those properties of geometrical objects that are expressed by the axioms may legitimately be used in a proof, so any particular interpretation of them in real terms is *in principle* irrelevant. In practice a good picture aids intuition, and it's a good bet that Hilbert's work on geometry was inspired by the usual mental picture of lines on a sheet of paper, not beer mugs on a table. He wasn't saying the traditional picture isn't useful, merely that it's not essential.

Beltrami's model was soon followed by others. A particularly beautiful one was devised by Henri Poincaré, who described a three-dimensional version thus:

Suppose, for example, a world enclosed in a large sphere and subject to the following laws: the temperature is not uniform; it is greatest at the centre, and gradually decreases as we move towards the circumference of the sphere, where it is absolute zero. The law of this temperature is as follows: if R be the radius of the sphere, and r the distance from the centre, then the absolute temperature will be proportional to $R^2 - r^2$. Suppose that in this world the linear dilation of any body is proportional to its absolute temperature. A moving object will become smaller and smaller as it approaches the circumference of the sphere. Although from the point of view of our ordinary geometry this world is finite, to its inhabitants it will appear infinite. As they approach the surface of the sphere they will become colder, and at the same time smaller and smaller. The steps they take are also smaller and smaller, so that they can never reach the boundary of the sphere. If to us geometry is only the study of the laws according to which rigid solids move, to these imaginary beings it will be the study of the laws of motion of solids *deformed by the differences in temperature* alluded to.

Poincaré goes on to elaborate this model so that the paths traced by light rays become curved—in fact, arcs of circles meeting the boundary at right angles—and obey non-Euclidean geometry. It is actually a three-dimensional model: the analogous model using a disc instead of a sphere produces a two-

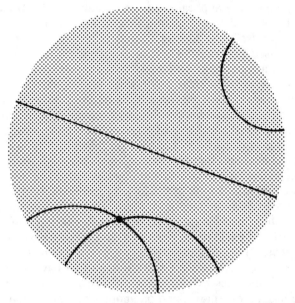

Fig. 5. Poincaré's model of two-dimensional hyperbolic geometry is a disc whose boundary is considered to be 'at infinity'. Straight lines in hyperbolic geometry are modelled as arcs of circles meeting the boundary at right angles.

dimensional non-Euclidean geometry. Felix Klein, who also devised models of non-Euclidean geometry, coined the name *hyperbolic* geometry for a model obeying Saccheri's hypothesis of the acute angle, and *elliptic* geometry for the obtuse angle. A model for elliptic geometry is very easy to find. Consider the surface of a sphere: interpret 'line' to mean 'great circle', and 'point' to mean 'pair of antipodal points'. (Great circles meet in two antipodal points: to obey Euclid's axiom that two lines meet in at most one point it is necessary to identify antipodal points in pairs. This is permissible by the Hilbert dictum about tables, chairs, and beer mugs.)

A matter of hindsight

To the modern mind it is astonishing how long it took for people to realize that geometries other than Euclid's are possible. The geometry of the sphere was already well developed, having important applications to astronomy, and hence navigation. Did nobody *notice* that in spherical geometry the parallel axiom is false? The answer is that they did; but they were unimpressed because the sphere is manifestly not the plane. Also, on a sphere any two lines, that is, great circles, meet at *two* points. This is clearly not what one would expect of a straight line. Only a person with a different logical point of view can get round this problem by pretending that antipodal points are the same! None of this would have been convincing at the time, because it was obvious that great circles aren't straight lines, and that two points aren't the same as one. To understand why it is convincing now, we must journey ahead in time and take a look at the logic of axiomatic systems.

The idea of an axiomatic system is that all assumptions are to be stated clearly at the start. In other words, the rules of the game are to be laid down precisely. You don't ask whether the axioms are 'true', any more than you ask whether the rules of chess are 'true'. However, you do expect to have a playable game. If the rules of chess were to insist on one page that white has the first move, and on another that black does, then it would be impossible even to start the game. In other words, we want the rules to be *consistent*. It is the same with axiomatic systems: the ultimate test is consistency, not truth. One way to prove that a system of axioms is consistent is to exhibit something that satisfies them. One example of a legal game of chess shows that the rules lead to a playable game; one *model* for an axiom system shows that its assumptions do not contradict themselves. For example, we can show that Euclidean geometry is consistent by taking the usual coordinate geometry model of the plane as a set of points (x, y) where x and y are numbers, and checking that all of the axioms hold. There is one fly in the ointment, however: the question of which ingredients are permissible. The coordinate model of geometry establishes consistency *only* if the raw materials—here arithmetic—are already known to be consistent. In other words, all consistency proofs are *relative*. Instead of asserting that system X is consistent, all we can show is that X is consistent provided some other system Y is. This viewpoint, now widely accepted, approaches the whole problem from a different direction. The original question was: 'Is Euclid's geometry the only true one?' But now even Euclidean geometry must come under scrutiny. Is *that* consistent? And the answer is that we have no idea. However, if arithmetic is consistent, then so is Euclid.

Now we ask: is *non*-Euclidean geometry consistent? And the answer is that again we don't know, but it is consistent provided either arithmetic, or Euclidean geometry, is. From the mathematical point of view, either Euclidean or non-Euclidean geometry is a reasonable thing to use. Which (if either) is the true geometry of physical space is revealed as a question for experiment, not mathematics. This seems so reasonable now that it's hard to appreciate how obscure it was then. The clarity is a matter of hindsight: without the unprecedented ideas of the nineteenth-century pioneers, everything would still be obscure.

Curved space

The discovery of non-Euclidean geometries freed mathematics from its previous rigid stance, and a host of strange geometries proliferated. One might imagine that these oddities, while of theoretical interest, are of no importance for the behaviour of real space. But that also turned out to be untrue. In 1854 a young German, Georg Bernhard Riemann, was due to deliver his qualifying lecture for the title of *Privatdozent*, which translates loosely as 'assistant lecturer'. The examiner was the great man himself, Gauss. Riemann offered several alternatives, some safe, some less so; and Gauss chose the most provocative, with the title 'On the Hypotheses which lie at the Foundation of Geometry'. Gauss later said he'd chosen that topic because he'd thought a lot about it himself, and wondered what the young man would have to say about it. Riemann, rising to the occasion, gave a brilliant lecture in which he imagined that the geometry of space might vary from point to point: hyperbolic here, Euclidean there, elliptic elsewhere. In two dimensions we can imagine this as the geometry of geodesics on some curved surface: the nature of the curvature determines the local geometry. Riemann extended this to multidimensional spaces, or *manifolds*. He ended prophetically by observing:

It remains to resolve the question of knowing in what measure and up to what point these hypotheses about manifolds are confirmed by experience. Either the reality which underlies space must form a discrete manifold or we must seek the ground of its metric relations outside it, in the binding forces which act on it. This leads us into the domain of another science, that of physics, into which the object of our work does not allow us to go today.

In 1870 William Kingdon Clifford took up the prophecy. 'I hold . . . that small portions of space are of a nature analogous to little hills on a surface which is on average flat . . . That this property of being curved or distorted is continually passed on from one portion of space to another after the manner of a wave.' And he suggested that all physical motion was really just the ripples of these waves. In Albert Einstein's *General Theory of Relativity*, published some sixty years after Riemann, gravitational force is viewed as a curvature of space-time. Near a large mass, space is more bent than it is a long way away. Particles moving through the space follow curved paths because of this geometric distortion. The best experiments yet devised all confirm Einstein's picture, which explains a number of observations that would not accord with a Euclidean geometry for space-time; for example, irregularities in the motion of the planet Mercury, and the bending of light by a star. More recently, astronomers have found that entire galaxies can act as 'gravitational lenses', focusing and distorting the light from distant sources. In one instance, what appeared to be two identical quasars close together in the sky were found to be a double image, in a distorted gravitational lens, of a *single* quasar on the far side of a massive galaxy. The galaxy was observed by 'subtracting' one image from the other. Still more recently, the study of solitary waves, or solitons, has suggested to some theorists that the behaviour of sub-atomic particles may resemble the moving hills imagined by Clifford, At any rate, we now believe that the geometry of space or space-time, on astronomical scales, is *not* Euclidean!

6

Sphereful symmetry

I compressed several fresh parcels of Pease in the same Pot, with a force equal to 1600, 800, and 400 pounds; in which Experiments, tho' the Pease dilated, yet they did not raise the lever, because what they increased in bulk was, by the great incumbent weight, pressed into the interstices of the Pease, which they adequately filled up, being thereby formed into pretty regular Dodecahedrons.

Stephen Hales

A pile of oranges outside a greengrocer's shop may not appear to hold any interest for the mathematician, beyond its obvious edibility. Appearances, as usual, are deceptive: the manner in which oranges are habitually stacked poses one of the oldest puzzles in mathematics, a problem that pre-dates Fermat's Last Theorem by three decades. It is a puzzle that relates not only to fruit displays, but to some of the deepest aspects of the structure of matter. It is known as Kepler's Sphere-Packing Problem, or the Kepler Problem for short. It has survived nearly four centuries without a scratch, but now a solution has been announced.

The mathematician who claims to have cracked it is Wu-Yi Hsiang, from the University of California at Berkeley; and his inspiration came from teaching a lecture course. The proof is currently being made available for checking by the mathematical community. As I write, some errors and gaps have emerged. It is not yet clear whether they are fatal. If the proof survives scrutiny, Hsiang will have achieved one of the most astonishing successes in the entire history of mathematics. Given the delays that occur in publishing, this chapter is something of a hostage to fortune; but the story is much too interesting to leave out just in case the proof falls apart between the manuscript going to the printer and the book appearing in print.

A present for my sponsor

Johannes Kepler achieved everlasting fame when he discovered that planets move in elliptical orbits, thereby paving the way for Isaac Newton's Law of Universal Gravitation and a large part of modern science. Kepler's imagination was wide-ranging, and his love of mathematical pattern, especially geometry, ran very deep indeed. We will see later how he tried to explain the spacing of the planets in terms of regular polyhedra. His sphere-packing problem arose from some deep thinking about snowflakes.

Snowflakes are crystals of ice, and their commonest form is a flat hexagon. Unlike the rather austere hexagons that decorate geometry texts, however, hexagonal snowflakes are decorated and embellished with beautiful fern-like structures, known to crystallographers as dendrites. No two snowflakes, it is said, are identical; but every one of them has hexagonal symmetry. Why? The question led Kepler to some remarkable insights into crystal structure, three centuries before physicists started to develop the atomic theory of matter. His ideas first saw print as a New Year's present to his sponsor. The title page of this unusual document reads:

JOHANN KEPLER

MATHEMATICIAN TO HIS IMPERIAL MAJESTY

A NEW YEAR'S GIFT

or

On the Six-Cornered Snowflake

Copyright licensed by His Imperial Majesty for fifteen years
Published by Godfrey Tampach at Frankfort on Main
in the year 1611

The volume is dedicated 'To the illustrious Counsellor at the Court of His Sacred Imperial Majesty, John Matthew Wacker of Wackenfels, Knight Bachelor, *etc.*, Patron of Men of Letters and of Philosophers, my Master and Benefactor.' Wacker was born in 1550 and died in 1619: he studied law at Strasburg and Geneva, and was ennobled in 1592. He had wide intellectual interests, especially in literature, and composed poetry, including an ode to a special local brew of beer. Colin Hardie, who translated Kepler's volume, remarks that this 'shows his taste for elegant trifles'. Wacker was thus the perfect recipient for Kepler's articulate and witty intellectual ramble.

In the opening pages Kepler explains that he was crossing a bridge, embarrassed by not having a suitable gift to offer his patron. 'Just then by a happy chance water-vapour was condensed by the cold into snow, and specks of down fell here and there on my coat, all with six corners and feathered radii. 'Pon my word, here was something smaller than any drop, yet with a pattern; here was the ideal New Year's Gift.' A short discussion leads him to the heart of the problem: is the form of the snowflake inherent in its internal structure, or is it imposed from outside?

The tightest pack

Kepler tackles the question with a breathtakingly rapid sequence of allusions and images—in which we can see the germs of several major ideas in modern crystallography—presented by way of homely examples such as bees making honeycombs. The key idea, says Kepler, is that of a space-filling solid: a polyhedron such that identical copies can be stacked together without leaving any gaps. He describes some of these, in particular a new one he has invented, a solid

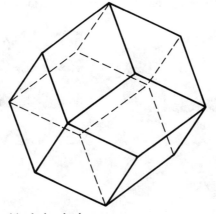

Fig. 6.   Rhombic dodecahedron.

built from twelve rhombic faces now known as the *rhombic dodecahedron*. This reminds him of something else: 'If one opens up a rather large-sized pomegranate, one will see most of its loculi squeezed into the same shape . . . What agent creates the rhomboid shape in the cells of the honeycomb and in the loculi of the pomegranate?' Clearly—at least in the pomegranate—the growth of the plant. 'But this is not a sufficient cause of its shape: for it is not from its formal properties that it induces this shape in the fruit, but it is assisted by material necessity. For the loculi to begin with, when they are small, are round, so long as there is enough space for them inside the rind. But later as the rind hardens, while the loculi continue to grow, they become packed and squeezed together . . .'

By now he is hot on the trail: the key to the honeycomb and pomegranate (and eventually also to the snowflake, and crystals in general) is the mathematical problem of packing identical spherical pellets into the smallest possible space. The associated space-filling solids are obtained by subsequently allowing these spheres to expand. Kepler finds that in the plane there are two ways to arrange the spheres: in a square lattice, like the squares of a chessboard, or in a hexagonal lattice, like the cells of a honeycomb. These arrangements in turn can be stacked in space in several ways. At first sight, there are four combinations; but Kepler observes that they reduce to three, of which one is by far the most interesting.

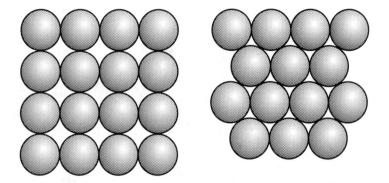

Fig. 7.   Two ways to arrange spheres in a plane layer.

'Square' layers can be stacked so that corresponding spheres are vertically above each other. The centres of the spheres lie at the corners of cubes which stack together to fill space in the obvious way. 'But', says Kepler, 'this will not be the tightest pack.' Alternatively (and this is what greengrocers do) the spheres in each layer can be laid so that they nestle into the gaps between four spheres in the layer below. With 'hexagonal' layers there are also these two possibilities, aligned or staggered; but staggered layers of hexagonally packed spheres lead to the *same* arrangement as staggered layers of squarely packed ones. The only difference is that the layers of one are tilted in comparison to the corresponding layers of the other. This is not so hard to see: use staggered 'square' packing to

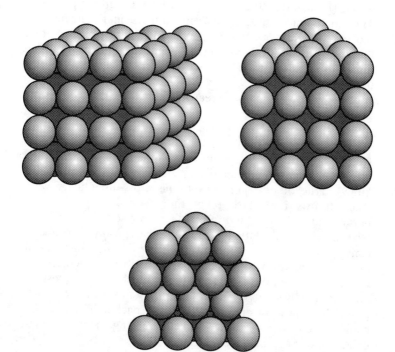

Fig. 8. Three ways to pack spheres in space by combining plane layers.

build a square pyramid of oranges, and then look at one of its sloping faces: you'll see one layer of a staggered 'hexagonal' packing. These three packings are known to crystallographers as the three-dimensional cubic lattice, face-centred cubic lattice, and hexagonal lattice.

If the spheres in these three arrangements are allowed to expand, like the loculi of pomegranates, then they acquire three characteristic shapes. For the cubic lattice, they become cubes; for the hexagonal lattice, hexagonal prisms; and for the face-centred cubic lattice, Kepler's beloved rhombic dodecahedra. Why do pomegranates choose to become rhombic dodecahedra? Because, says Kepler, to begin with their seeds can jiggle around as they grow, so they try not to waste any space. That leads to the face-centred cubic lattice, for which (he asserts) 'the packing will be the tightest possible'. This single innocuous phrase dropped mathematics neatly into the mire of Kepler's Sphere-Packing Problem: to *prove* that the most economical packing of identical spheres in three dimensional space is the face-centred cubic lattice—the one used by greengrocers.

One seven-word phrase created mathematical havoc lasting (at least) 380 years.

Proof of the pomegranate

Kepler's phrase has passed into anecdotal history as the statement that 'most mathematicians believe, and all physicists know'. True or not, fair or not, this cross-disciplinary wisecrack throws into stark relief the distinction between experimental evidence and proof. Can we demonstrate that face-centred cubic lattice packing is a logically inescapable consequence of most efficient packing? Until Hsiang came along, the answer was 'no'. Indeed, until 1910, mathematicians would have given the same answer to a much simpler question, the two-dimensional version of Kepler's Problem. This is to prove that the most efficient method of packing circles in the plane is the hexagonal lattice—like the arrangement of the red balls at the start of a snooker game.

Before we investigate the nature of the difficulty, we'd better agree on what we mean by 'efficient' packing. Mathematically, the problem is posed for packings that fill the whole of space, and not just a limited region. Anyone who buys groceries knows that the best way to pack oranges into a box or basket of one particular shape may not be the best way to pack them into another one. These 'edge effects' lead to incredible complications; for instance, hardly anything worthwhile is known about the most efficient way of packing spheres into ordinary box-shaped boxes (technically known as cuboids, or more fancily as rectangular parallelepipeds).

The efficiency of a packing is measured by a number, its *density*, which basically is the proportion of space that is filled with spheres. Now, with an infinite volume of space available, and an infinite number of spheres, it's important not to calculate this proportion as ∞/∞. That way lies disaster. Somehow, two finite volumes have to get in on the act. The solution is to define the density by considering a large but finite region of space, calculating what proportion is filled by spheres, and then letting the size of the region tend to infinity. Edge effects do still raise their ugly heads, for any particular region along the way, but their contribution to the answer diminishes rapidly as the region becomes bigger, and in the limit edge-effects play no role whatsoever.

Let's see what the results look like. In the plane, Kepler distinguishes two types of packing: the square lattice and the hexagonal. Their densities are $\pi/4$ and $\pi/2\sqrt{3}$ respectively, or about 0.7854 and 0.9069. Clearly the hexagonal lattice has the greater density. In three-dimensional space, the cubic lattice has density $\pi/6 = 0.5236$, the hexagonal lattice $\pi/3\sqrt{3} = 0.6046$, and the face-centred cubic lattice $\pi/3\sqrt{2} = 0.7404$.

It's easy enough to tell which is biggest, so what's the problem?

Well . . . Kepler wasn't talking about just these particular packings. Of course, their densities can be calculated and compared, and we can see which one wins, but that's not the question. The question is, might there be some *other* packing, possibly very complicated, maybe even 'random', with greater density? Like the heroine of a Victorian melodrama strapped to Dr Frankenstein's operating table with enormous electrodes attached to her head and sparks crackling round the walls, we realize the true horror of our predicament. We are being asked (gulp) to consider *every conceivable packing* of the whole of infinite space by infinitely many spheres, however intricately arranged.

We desperately need a hero, someone who can short-circuit the entire mess.

Vegetable Staticks

Densest packings must be incompressible; that is, it must be impossible to move the spheres around to pack them closer together. Otherwise, they wouldn't be densest. Conversely, might every incompressible packing be maximally dense? If so, that would open up new lines of attack. However, simple experiments strongly suggest that there are huge numbers of incompressible packings whose densities are far too small. These are 'random' packings, obtained by adding spheres one at a time in no particular pattern—like pouring lead shot into a bowl. Even though it does not amount to rigorous proof, experimental evidence is not without value. It can motivate the hero by convincing him he's on the right track. Our chapter quote is taken from Stephen Hales's bizarrely titled *Vegetable Staticks* of 1727. It is questionable whether he actually saw what he claims, because *regular* dodecahedra, unlike their rhombic cousins, don't pack space; but in random packings it's not unusual to find good approximations to regular dodecahedra, so Hales may just have been selective. In 1939 the botanists J. W. Marvin and E. B. Matzke experimented using lead shot in a steel cylinder with a plunger. They verified that if the shot were stacked like oranges and compressed then they became rhombic dodecahedra—oddly, one of the facts that the mathematicians *could* prove, and which therefore did not require experimental confirmation! They observed that random packing led to 'irregular 14-faced bodies' rather than Kepler's rhombic dodecahedra. In 1959 J. D. Bernal compressed spheres of plasticene rolled in powdered chalk, and obtained 13.3 as the average number of faces. Dodecahedra, be they rhombic or regular, were conspicuous by their absence; but in none of these experiments was much care taken to ensure efficient packing. G. D. Scott experimented with ball-bearings, shaking the container as it was filled, and found that such 'random' packings attained a density no larger than 0.6366, which is not as great as that for the face-centred cubic lattice.

Built on sand

Because of the way they are prepared, random packings are incompressible: squashing the spheres does not cause any movement. They are what mathematicians would call *local* maxima of the density, meaning that any *small* change in the positions of the spheres causes the density to decrease. But Kepler was asking for a *global* maximum: an arrangement whose density cannot be increased even by large changes.

There is an amusing consequence of local maximality. Let me ask a question. When you walk on sand, is it compressed by your weight?

Silly question? I admit it looks that way, but ... If the sand is in an incompressible random packing, which is highly plausible, then by definition it *can't* be compressed any further. If any particles of sand move, then the space between them must *increase*. In 1885 Osborne Reynolds pointed this out to the British Association at Aberdeen: 'as the foot presses upon the sand when the falling tide leaves it firm, that portion of it surrounding the foot becomes momentarily dry.' He explained that the pressure of the foot, far from compressing the sand beneath it, actually causes it to *dilate*, opening up new gaps for water to flow in from the surrounding sand. 'On raising the foot we generally see that the sand under and around it becomes wet for a little time. This is because the sand contracts when the distorting forces are removed, and the excess of water escapes at the surface.' This tale contains the valuable warning that sphere-packings are often counter-intuitive. As Lord Kelvin remarked in 1904: 'Of all the two hundred thousand men, women, and children who, from the beginning of the world, have ever walked on wet sand, how many, prior to the British Association meeting at Aberdeen in 1885, if asked "is the sand compressed under your foot" would have answered otherwise than "Yes!"?'

Experiment, then suggests that random arrangements, even though they are incompressible, do not even come close to the densest packing possible. This certainly strengthens our belief that there may exist a mathematical proof that Kepler was right—but it offers no clue as to what it might look like. Indeed, by revealing the probable existence of huge numbers of 'spurious' packings with *local* density maxima, it shows just how nasty the problem is likely to be. For example, an attractive approach to a proof would be to show that in any packing other than Kepler's a few spheres can always be rearranged to improve the density. But the existence of local maxima prevents any such approach from succeeding.

Lattice packings

Positive results did arrive—but desperately slowly. In 1831 Carl Friedrich Gauss introduced the concept of a lattice—a regular grid-like arrangement of points in the plane or in space—and related it to previous work in number theory by Joseph-Louis Lagrange. Define a *lattice packing* to be one for which the centres of the spheres form a lattice. By exploiting the number-theoretic connection and applying some clever results of Lagrange, Gauss proved that the hexagonal lattice is the densest *lattice* packing in the plane. Soon after, L. A. Seeber wrote a book on number theory, proved a system of inequalities satisfied by things called 'ternary reduced forms', and conjectured a result equivalent to the face-centred cubic lattice being the densest *lattice* packing in three-dimensional space. Gauss, in a review of Seeber's book, deduced the conjecture from the inequalities and thereby proved the theorem. The theory of lattice packings took off: today it has applications not just to number theory, but to communications and coding. Even though David Hilbert included it in his famous 1900 list of 23 major unsolved problems, Kepler's Problem on *non*-lattice packings languished.

It was partially revived in 1892, when Axel Thue lectured to the Scandinavian Natural Science Congress, outlining a *solution* of the two-dimensional version of the Kepler Problem: a proof that the densest packing of equal circles in the plane, regular or not, is the hexagonal lattice. The published version of his lecture is sketchy and it seems difficult to reconstruct (and hence check the validity of) his proof. In 1910 he gave another proof from a very different viewpoint, which appears fairly convincing, except perhaps for some technical points which may be tricky but which he takes for granted. In 1940 L. Fejes Tóth found a different proof, avoiding these snags, and it was followed almost immediately by further proofs from B. Segre and Kurt Mahler.

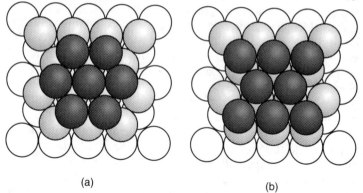

Fig. 9. Different ways to stack layers. (a) leads to a non-lattice packing in which spheres in the third layer are directly above those in the first; (b) leads to a lattice packing. Both packings have the same density.

(a)

(b)

Their methods, unfortunately, offered no clue about the three-dimensional version. Moreover, there were some warning signs that Kepler might even be wrong. The first was very likely known to Kepler himself. We've seen that a major obstacle in the way of a proof is the possibility of efficient non-lattice packings. Despite the experimental evidence that the density of random incompressible packings is too low, Kepler's conjecture is certainly very delicate, because there definitely exist non-lattice packings that are *precisely* as dense as the face-centred cubic lattice. They are easily described. Think of the face-centred cubic lattice as being formed out of layers, each a hexagonal lattice in the plane. This is Kepler's second interpretation of the same packing pattern. Start with the first layer, then add a second whose spheres fall between those of the first. At the third layer, there are two distinct ways to do this. In one of them, the spheres in the new layer sit directly above the corresponding spheres in the first. In the other, they do not. The second arrangement leads to the face-centred cubic lattice. The first, if repeated, leads to a non-lattice pattern. It is clear from the way the layers fit together, however, that the density is the same for either arrangement.

Moreover, as W. Barlow pointed out in 1883, we can continue to pile hexagonal layers on top of each other, at each stage selecting one or other of these two possibilites, at random. So we would obtain a packing which is 'random' in the vertical direction, though regular in the two horizontal directions, whose density is exactly the same as that of the face-centred cubic lattice.

This certainly suggests that it is far from obvious that the best packing—or more accurately, at least one packing among those with the greatest density—must live on a lattice. Indeed, if we extend the problem of sphere-packing into higher dimensions (packing hyperspheres in hyperspace), then for spaces of very high dimension—around 1000—the densest known packings are *non*-lattice. In a thousand-dimensional space, matter may not have to form itself into crystal lattices—so to speak. Thus the lattice nature of the best packing is both delicate and questionable. This all makes the mathematician's life a lot harder, and tends to sap the resolution of would-be melodramatic heros.

In 1958 C. A. Rogers proved that the maximum density for any packing in three dimensional space is at most 0.7797, not *too* much greater than the conjectured value of 0.7404. His bound was subsequently improved to 0.77844 in 1986, and 0.77836 in 1988. Like Olympic athletes, mathematicians were being forced to introduce extra decimal places in order to record any progress at all. Despite strenuous efforts, sometimes intense, sometimes sporadic, over more than three centuries, the entire enterprise was grinding to a halt. The situation, as John Milnor had said in 1974, was 'scandalous'.

With one bound, our hero was free

Wu-Yi Hsiang's main area of research is a rather complicated but beautiful part of topology. Like many topologists, he has excellent visual intution, and in the spring of 1990 he agreed to create a new course at Berkeley—on a totally outdated branch of mathematics: classical geometry. The image of Victorian melodrama is not as strained as it may have appeared. To flex his mental muscles our hero took a hard look at the toughest problem in the area that he could find, which was the Kepler problem. He has now claimed a solution. Hsiang is quoted as saying 'I got hooked on it. The more I thought about this problem, the more beautiful it appeared.'

Fittingly, his proof is also in the classical vein. In principle it could have been discovered long ago. It does *not* use any of the sophisticated machinery developed by mathematicians over the last century or so. It is cast entirely in the classical language of spherical geometry, vectors, and calculus. Why was it not found before? Hsiang's view is that we just haven't understood spherical geometry very well. The proof does require a level of mathematical maturity that is more appropriate to the twentieth century; and in particular, it runs to some hundred typed pages.

The central idea is a result known as the 'local density conjecture'. As with Kepler's pomegranates, imagine each sphere swelling at a uniform rate until it touches its neighbours along flat interfaces. Then each sphere is surrounded by a polyhedron, its *local cell*. The *local density* of any particular sphere is the ratio of its volume to that of its local cell. If any upper bound for—that is, fixed number greater than—the local density can be found, it will *automatically* also be an upper bound for the overall density of the entire packing.

It had been suspected for a long time that the maximum local density is 0.7547, or more precisely

$$\frac{\pi\sqrt{5+\sqrt{5}}}{5\sqrt{10}(\sqrt{5}-2)},$$

attained when the local cell is a regular dodecahedron. Proving this conjecture *alone* would improve the bound on packing density from 0.77836, the best value known by 1988, to 0.7547, much closer to the density of 0.7404 conjectured by Kepler. Moreover, the local density conjecture looks much more accessible than the full Kepler Problem, because it deals not with infinitely many spheres, but just those few that surround a given one. All mathematicians absorb with their mothers' milk the principle that local problems are almost always simpler than global ones.

Mind you, nobody could prove the local density conjecture either.

So Hsiang started with that. 'The central issue', he states, 'is how to achieve effective lower bound estimations of the volume of a local cell or that of a cluster of local cells.' He found a new method for estimating the volume of local cells, by cutting them up—not into cones, as everyone else had tried, but double cones. According to Hsiang, this idea exploits the spherical geometry of local cells more effectively. With the aid of a concept of 'semiglobal' density, he then uses the local version of the problem as a stepping-stone to the global version.

One hundred pages of tricky geometry . . . clever it certainly is, but is the proof *correct*? As I write, some flaws are beginning to emerge, both in the proof of the local density conjecture and in its application to the full problem. It may take some time before the status of his ideas is resolved. Errors can often be put right: gaps can be patched. On the other hand, flaws can also prove fatal. Maybe we'll have to wait another 380 years . . .

The kissing number

Hsiang claims that his ideas can also solve another problem about spheres, this time in four dimensions. In a multi-dimensional space, how many equal spheres can touch a single sphere of the same size, without intersecting each other? The answer is called the *kissing number* in that many dimensions. Think about it in two dimensions first, where the 'spheres' are

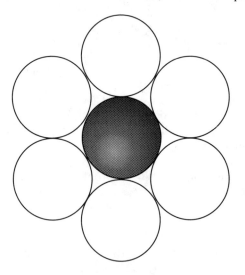

Fig. 10.   The kissing number of a circle is 6.

The kissing number 83

circular discs, which we can realize using coins. Place a coin on the table. Now place as many other coins, all the same size as the first, so that they all touch it. What's the largest number that will fit? A little experiment makes the answer clear: *six*. Six discs will fit exactly, forming a regular hexagon.

The general kissing problem is just like that, but with spheres. For the three-dimensional case, imagine ball-bearings rather than coins, arranged in three dimensions around a single central ball of the same size. Isaac Newton and David Gregory argued about the problem in 1694. Newton said the answer was 12, Gregory thought 13 might be possible. Twelve is certainly possible: place the surrounding spheres at the vertices of a regular icosahedron. But there's room to move them around, so maybe an extra sphere can be crammed in somehow. However, in the nineteenth century C. Bender, R. Hoppe, and S. Günther offered proofs that Newton was right. A detailed proof was published in 1953 by K. Schütte and B. L. van der Waerden. The best-known proof was found by John Leech in 1956, and even that proof is far from easy. One

TABLE 6.1. *Kissing numbers*

Dimension	Kissing number	Dimension	Kissing number
1	**2**	13	1,130–2,233
2	**6**	14	1,582–3,492
3	**12**	15	2,564–5,431
4	**24***	16	4,320–8,313
5	40–46	17	5,346–12,215
6	72–82	18	7,398–17,877
7	126–140	19	10,668–25,901
8	**240**	20	17,400–37,974
9	306–380	21	27,720–56,852
10	500–595	22	49,896–86,537
11	582–915	23	93,150–128,096
12	840–1,416	24	**196,560**

Notes: The values show the range between the largest values yet achieved and the smallest upper limits yet established: those in bold are the final answers since the two limits are equal.
* If Hsiang's proof is verified; 24–25 if not.

source of difficulty is that the arrangement isn't rigid: there's a lot of freedom to slide spheres around. In fact you can rearrange the twelve spheres in any way you like by sliding them without intersecting, always touching the central sphere. So they don't *have* to be at the vertices of an icosahedron.

The known results for dimensions up to 24 are shown in Table 6.1 (page 83).

The exact number is definitely known for 1, 2, 3, 8, and 24 dimensions. Hsiang has just announced that he has solved the problem in four dimensions, and there the correct kissing number is 24. It may seem rather curious that the answer is *unknown* in five dimensions, but known in eight. And in *twenty-four*-dimensional space, for heaven's sake. Those results were found by Andrew Odlyzko and Neil Sloane. They found a way to estimate a good upper bound, and it was so good it turned out to be the same as the best known arrangement of spheres. So that was that. The twenty-four-dimensional arrangement is given by the Leech lattice, whose symmetries form the Conway simple group. There's something rather unusual about space of eight and twenty-four dimensions. All sorts of sphere-packing and lattice problems work out nicest in those dimensions. In particular, in eight and twenty-four dimensions the arrangements of spheres are rigid: answers to the kissing problem are unique. It's aways easier to pin down an answer when there's only one of it.

What about the snowflakes?

Oh, yes. Snowflakes. What with all this sphere-packing, Kepler's original motivation has got a bit lost. Time to remedy that.

What Kepler deduces from his experiments on pomegranates is that when large numbers of identical tiny particles assemble themselves in the most efficient manner possible, they automatically create geometric order: squares and hexagons, lattices, atomic piles—of oranges. He goes on to lay down what is effectively an atomic theory of crystal structure, very similar in spirit to the one that we recognize today. He traces the shapes of all crystals, snowflakes included, to some internal 'formative faculty' of the material from which they are made. He concludes with these words: 'Now that I have knocked at the door of chemistry and see how much remains to be said before we can get hold of our cause, I prefer to hear what a man of your great acumen thinks rather than to tire myself with further discourse.'

Over the years, Kepler's theories were developed, often independently, by many other scientists: Robert Boyle, Robert Hooke, Abraham Werner, Jean Romé de l'Isle, Christian Huygens, René Just Haüy, Christian Weiss, Friedrich Mohs, Auguste Bravais, Evgraf Fedorov, and Arthur Schoenflies. Despite this galaxy of scientific stardom, not until 1915 was it demonstrated experimentally that crystals really are formed from identical particles—atoms—arranged in regular lattices. In that year Lawrence Bragg achieved such a demonstration using X-ray diffraction, and the science of crystallography changed forever. Kepler was three centuries ahead of his time.

Snowflakes, it is now thought, begin life as a tiny hexagonal 'seed' created by such a lattice packing of atoms. As they fall and rise on winds, high in the atmosphere, the conditions (temperature, humidity, pressure) around them change, and so do the pattern of growth. But the crystals are so small that at any given instant the conditions on all six sides are the same; they vary only in time, not in space. So the same growth pattern occurs along each of the six sides, preserving hexagonal symmetry. The atmospheric conditions themselves, however, can vary almost randomly, creating a vast range of distinct shapes.

Even now, we have not reached the end of the story. Physicists would like very much to deduce the observed properties of bulk matter from quantum mechanics. That is, they would like to write down the quantum-mechanical equations for a vast number of interacting atoms, and prove that these have solutions that correspond to the properties of gases, liquids, solids, and crystals. Success as yet is limited: it has been proved that gases exist. Liquids, solids, and crystals remain far out of reach.

A solution of Kepler's problem would not prove that the existence of crystals is a logical consequence of quantum mechanics. It would, however, provide a purely geometric proof that when spherical particles pack themselves with maximum density, then a crystalline structure must result.

However, if it were possible to deduce from quantum mechanics that the atoms in a crystal must pack themselves together with maximum density, then Hsiang's (claimed) theorem—or his techniques—could come into play. Many lattices other than the face-centred cubic occur in crystals, of course; but most crystals involve several distinct atoms, and would correspond to different packing problems. The Kepler Problem is just a prototype. Pure speculation, I admit; but even if I'm wrong, a proof that Kepler was right would bring the whole subject full circle, providing a solid mathematical basis that links the geometric structure of crystals to the atomic nature of matter. That would be an amazing achievement, especially if it can be based on a New Year's present to a sponsor and a course on out-of-date mathematics.

7

The miraculous jar

To see a world in a grain of sand,
And a Heaven in a wild flower,
Hold Infinity in the palm of your hand,
And Eternity in an hour.

William Blake

Pick up any mathematics book or journal and you probably won't get more than a page or two into it before you bump headlong against infinity. For example, selecting a few books from my shelf at random, the second *line* of the introduction to Harold Davenport's *The Higher Arithmetic* refers to 'the natural numbers 1, 2, 3, . . .' and the sequence of dots is intended to imply that the natural numbers are infinite in extent. Page 2 of *Numerical Analysis* by Lee Johnson and Dean Riess quotes the infinite series for the exponential function. B. A. F. Wehrfritz's *Infinite Linear Groups*—need I say more? Page 1 of *Nonlinear Dynamics and Turbulence*, edited by G. I. Barenblatt, G. Iooss, and D. D. Joseph, refers to 'the Navier-Stokes equations or a finite-dimensional Galerkin approximation', correctly leading us to infer that mathematicians treat the full Navier-Stokes equations as an infinite-dimensional object. Page 434 of *Winning Ways* by Elwyn Berlekamp, John Conway, and Richard Guy talks of a game whose position is '∞, 0, ± 1, ± 4', where ∞ is the standard symbol for 'infinity'. You may feel that page 434 is stretching the point a bit, but it's the sixth page in volume 2, and I didn't check volume 1.

Infinity is, according to Philip Davis and Reuben Hersh, the 'Miraculous Jar of Mathematics'. It is miraculous because its contents are inexhaustible. Remove one object from an infinite jar and you have, not one fewer, but *exactly the same number* left. It is paradoxes like this that forced our forefathers to be wary of arguments involving appeals to the infinite. But the lure of infinity is too great. It is such a marvellous place to lose awkward things in. The number of mathematical proofs that succeed by pushing everything difficult out to infinity, and watching it vanish altogether, is itself almost infinite. But what do we really mean by infinity? Is it wild nonsense, or can it be tamed? Are the infinities of mathematics real, or are they clever fakes, the wolf of the infinite in finite sheep's clothing?

Hilbert's hotel

If you are laying a table and each place-setting has one knife and one fork, then you know that there are just as many knives as forks. This is true whether you lay out an intimate candlelit dinner for two or a Chinese banquet for 2,000, and you don't need to know how many places are set to be sure the numbers agree. This observation is the cornerstone of the number concept. Two sets of objects are said to be in *one-to-one correspondence* if to each object in one there corresponds a unique object in the other, and vice versa. Sets that can be placed in one-to-one correspondence contain the same number of objects.

When the sets are infinite, however, paradoxes arise. For example, Hilbert described an imaginary hotel with infinitely many rooms, numbered 1, 2, 3, . . . One evening, when the hotel is completely full, a solitary guest arrives seeking lodging. The resourceful hotel manager moves each guest up a room, so that the inhabitant of room 1 moves to room 2, room 2 to 3, and so on. With all guests relocated, room 1 becomes free for the new arrival! Next day an Infinity Tours coach arrives, containing infinitely many new guests. This time the manager moves the inhabitant of room 1 to room 2, room 2 to 4, room 3 to 6, . . . , room n to $2n$. This frees all odd-numbered rooms, so coach passenger number 1 can go into room 1, number 2 to room 3, number 3 to room 5, and, in general, number n to room $2n - 1$. Even if infinitely many infinite coachloads of tourists arrive, everybody can be accommodated.

Similar paradoxes have been noted throughout history. Proclus, who wrote commentaries on Euclid in about AD 450, noted that the diameter of a circle divides it into two halves, so there must be twice as many halves as diameters. Philosophers in the Middle Ages realized that two concentric circles can be matched one-to-one by making points on the same radius correspond; so a small circle has just as many points as a large one. In Galileo's *Mathematical Discourses and Demonstrations* the sagacious Salviati raises the same problem: 'If I ask how many are the Numbers Square, you can answer me truly, that they be as many as are their propper roots; since every Square hath its Root, and every Root its Square, nor hath any Square more than one sole Root, or any Root more than one sole Square.' To this the seldom-satisfied Simplicius replies: 'What is to be resolved on this occasion?' And Salviati cops out with: 'I see no other decision that it may admit, but to say, that all Numbers are infinite; Squares are infinite; and that neither is the multitude of Squares less than all Numbers, nor this greater than that: and in conclusion, that the Attributes of Equality, Majority, and Minority have no place in Infinities, but only in terminate quantities.'

Infinity in disguise

Galileo's answer to the paradoxes is that infinity behaves differently from anything else, and is best avoided. But sometimes it's very hard to avoid it. The problem of infinity arose more insistently in the early development of calculus, with the occurrence of infinite series. For example, what is

$$1 + 1/2 + 1/4 + 1/8 + 1/16 + \ldots ?$$

It's easy to see that as the number of terms increases, the sum gets closer and closer to 2. So it's attractive to say that the whole infinite sum is *exactly* 2. Newton made infinite series the foundation of his methods for differentiating and integrating functions. So the problem of making sense of them must be faced. And infinite series are themselves paradoxical. For example, what does the series

$$1 - 1 + 1 - 1 + 1 - 1 + \ldots$$

add up to? Written like this

$$(1 - 1) + (1 - 1) + (1 - 1) + \ldots$$

the sum is clearly 0. On the other hand,

$$1 - (1 - 1) - (1 - 1) - \ldots$$

is clearly 1. So $0 = 1$, and the whole of mathematics collapses in a contradiction.

Calculus was much too important for its practitioners to be put off by minor snags and philosophical problems like this. Eventually the matter was settled by reducing statements about infinite sums to more convoluted ones about finite sums. Instead of talking about an infinite sum $a_0 + a_1 + a_2 + \ldots$ having value a, we say that the *finite* sum $a_0 + a_1 + \ldots + a_n$ can be made to differ from a by less than any assigned error ε, provided n is taken larger than some N (depending on ε). Only if such an a exists does the series *converge*, that is, is the sum considered to make sense. In the same way the statement 'there are infinitely many integers' can be replaced by the finite version 'given any integer, there exists a larger one'. As Gauss put it in 1831: 'I protest against the use of an infinite quantity as an actual entity; this is never allowed in mathematics. The infinite is only a manner of speaking, in which one properly speaks of limits to which certain ratios can come as near as desired, while others are permitted to increase without bound.' Today, in any university course on analysis, students are taught to handle the infinite in this way. A typical problem might be: 'prove that $(n^2 + n)/n^2$ tends to 1 as n tends to infinity'. Woe betide the student who answers '$(\infty^2 + \infty)/\infty^2 = \infty/\infty = 1$'. But also woe betide him who writes '$(n^2 + n)/n^2 = 1 + 1/n$, now let $n = \infty$ to get $1 + 1/\infty = 1 + 0 = 1$', although this is arguably correct. (Before you're allowed to write sloppy things like that you must prove your mathematical machismo by going through the tortuous circumlocutions needed to make it unobjectionable. Once you've learned, the hard way, not to be sloppy, nobody minds if you are!)

This point of view goes back to Aristotle, and is described as *potential infinity*. We do not assert that an actual infinite exists, but we rephrase our assertion in a form that permits quantities to be as large as is necessary at the time. No longer do we see the miraculous jar as containing a true infinity of objects; we just observe that however many we take out, there's always another one in there. Put that way, it sounds like a pretty dubious distinction; but on a philosophical level it avoids the sticky question: 'How much stuff is there in that jar?'

Sum crisis!

However, there were still bold souls who continued to play about with the idea of 'actual' infinity; to think of an infinite set not as a process 1, 2, 3, ... which could in principle be continued beyond any chosen point, but as a completed, infinite whole. One of the first was Bernard Bolzano, who wrote a book called *Paradoxes of the Infinite* in 1851. But Bolzano's main interest was in providing solid foundations for calculus, and he decided that actually infinite sets aren't really needed there.

In the late nineteenth century there was something of a crisis in mathematics. Not fancy philosophical paradoxes about infinity, but a solid down-to-earth crisis that affected the day-to-day technique of working mathematicians, in the theory of Fourier series. A Fourier series looks something like this:

$$f(x) = \cos x + \frac{1}{2} \cos 2x + \frac{1}{3} \cos 3x + \ldots$$

and was developed by Joseph Fourier in his work on heat flow. The question is, when does such a series have a sum? Different mathematicians were obtaining contradictory answers. The whole thing was a dreadful mess, because too many workers had substituted plausible 'physical' arguments for good logical mathematics. It needed sorting out, urgently. Basically the answer is that a Fourier series makes good sense provided the set of values x, at which the function f behaves badly, is not itself too nasty. Mathematicians were forced to look at the fine structure of sets of points on the real line. In 1874 this problem led Georg Cantor to develop a theory of *actually* infinite sets, a topic that he developed over the succeeding years. His brilliantly original ideas attracted attention and some admiration, but his more conservatively minded contemporaries made little attempt to conceal their distaste. Cantor did two things. He founded Set Theory (without which today's mathematicians would find themselves tongue-tied, so basic a language has it become), and he discovered in so doing that some infinities are bigger than others.

Cantor's paradise

Cantor started by making a virtue out of what everyone else had regarded as a vice, He *defined* a set to be infinite if it can be put in one-to-one correspondence with a proper part (subset) of itself. Two sets are equivalent or have the same *cardinal* if they can be put in one-to-one correspondence with each other. The smallest infinite set is that comprising the natural numbers $\{0, 1, 2, 3, \ldots\}$. Its cardinal is denoted by the symbol $\aleph_0$ (aleph-zero) and this is the smallest infinite number. It has all sorts of weird properties, such as

$$\aleph_0 + 1 = \aleph_0, \ \aleph_0 + \aleph_0 = \aleph_0, \ \aleph_0^2 = \aleph_0$$

but nevertheless it leads to a consistent version of arithmetic for infinite numbers. (What do you expect infinity to do when you double it, anyway?) Any set with cardinal $\aleph_0$ is said to be *countable*. Examples include the sets of negative integers, all integers, even numbers, odd numbers, squares, cubes, primes, and—more surprisingly—rationals. We are used to the idea that there are far more rationals than integers, because the integers have large gaps between them whereas the rationals are densely distributed. But that intuition is misleading because it forgets that one-to-one correspondences don't have to respect the order in which points occur. A rational p/q is defined by a pair (p, q) of integers, so the number of rationals is $\aleph_0^2$. But this is just $\aleph_0$ as we've seen.

After a certain amount of this sort of thing, one starts to wonder whether *every* infinite set is countable. Maybe Salviati was right, and $\aleph_0$ is just a fancy symbol for ∞. Cantor showed this isn't true. The set of real numbers is uncountable. There is an infinity bigger than the infinity of natural numbers! The proof is highly original. Roughly, the idea is to assume that the reals are countable, and argue for a contradiction. List them out, as decimal expansions. Form a new decimal whose first digit after the decimal point is different from that of the first on the list; whose second digit differs from that of the second on the list; and in general whose nth digit differs from that of the nth on the list. Then this new number cannot be anywhere in the list, which is absurd since the list was assumed to be complete. This is Cantor's 'diagonal argument', and it has cropped up ever since in all sorts of important problems. Building on this, Cantor was able to give a dramatic proof that transcendental numbers must exist. Recall the immense difficulties that had been encountered in this problem. Cantor showed that the set of algebraic numbers is countable. Since the full set of reals is uncountable, there must exist numbers that are not algebraic. End of proof (which is basically a triviality); collapse of audience in incredulity. In fact Cantor's argument shows more: it shows that there must be uncountably many transcendentals! There are *more* transcendental numbers than algebraic ones; and you can prove it without ever exhibiting a single example of either. It must have seemed like magic, not mathematics.

Even Cantor had his moments of disbelief. When, after three years of trying to demonstrate the opposite, he proved that *n*-dimensional space has exactly the same number of points as 1-dimensional space, he wrote: 'I see it but I do not believe it.' Others felt a little more strongly, for example Paul du Bois-Reymond: 'It appears repugnant to common sense.' There were also some paradoxes whose resolution did not just require imaginative development of a new but consistent intuition. For example, Cantor showed that, given any infinite cardinal, there is a larger one. There are infinitely many different infinities. But now consider the cardinal of the set of all cardinals. This must be larger than any cardinal whatsoever, including itself! This problem was eventually resolved by restricting the concept of 'set', but I wouldn't say people are totally happy about that answer even today.

Mathematicians were divided on the importance of Cantor's ideas. Leopold Kronecker attacked them publicly and vociferously for a decade; at one point Cantor had a nervous breakdown. But Kronecker had a very restrictive philosophy of mathematics—'God made the integers, all else is the work of Man'—and he was no more likely to approve of Cantor's theories than the Republican Party is likely to turn the Mid-West over to collective farming. Poincaré said that later generations would regard them as 'a disease from which one has recovered'. Hermann Weyl opined that Cantor's infinity of infinities was 'a fog on a fog'. On the other hand Adolf Hurwitz and Hadamard discovered important applications of

Set Theory to analysis, and talked about them at prestigious international conferences. Hilbert, the leading mathematician of his age, said in 1926: 'No one shall expel us from the paradise which Cantor created', and praised his ideas as 'the most astonishing product of mathematical thought'. As with other strikingly original ideas, only those who were prepared to make an effort to understand and *use* them in their own work came to appreciate them. The commentators on the sidelines, smugly negative, let their sense of self-importance override their imagination and taste. Today the fruits of Cantor's labours form the basis of the whole of mathematics.

The continuum hypothesis

After Cantor's insight, mathematicians rapidly came to realize that there are three fundamentally different types of domain in which they can operate: finite, countable, and uncountable. Styles of argument are available in the finite case that cannot be used anywhere else: for example the 'pigeonhole principle' that you can't put $n + 1$ objects into n boxes with at most one in each. Hilbert's hotel shows that this is untrue when $n = \aleph_0$, and the same goes for any infinite n. Countable sets again have their own special features; there are plenty of occasions where $1, 2, 3, \ldots$ is simpler than $1, 2, 3, \ldots, n$. For example

$$1 + \frac{1}{4} + \frac{1}{9} + \ldots = \pi^2/6$$

exactly, whereas nobody has a formula for

$$1 + \frac{1}{4} + \frac{1}{9} + \ldots + 1/n^2.$$

The reason that $1, 2, 3, \ldots$ is simpler than $1, 2, 3, \ldots, n$ is plain: you don't have to worry about the n on the end! On the other hand, by counting $1, 2, 3, \ldots$ you eventually get to anything you want, so you can approach a countable set by way of a series of finite ones; whereas for the uncountable case nothing like this is possible. It's a very coarse but very important distinction; and you can tell a lot about a mathematician's likes and dislikes by asking him what size of set he feels happiest working with.

Cantor proved that the cardinal of the reals is larger than $\aleph_0$, but he left unresolved the question: is there anything in between? This problem became known as the *Continuum Hypothesis*. Every attempt to prove it failed dismally; but so did every attempt to construct a set with more elements than the integers but fewer than the reals. (Does this remind you of something? Read on.) Only in 1963 did Paul Cohen prove that the answer depends on your version of set theory. There are Cantorian set theories in which it is true; and non-Cantorian set theories in which it is false. It's Euclid's parallel axiom all over again. And we all ought to feel a lot less self-satisfied if we observe that it took modern mathematicians just as long to cotton on to the existence of non-Cantorian set theory as it did our predecessors to stumble across non-Euclidean geometry. 'The only lesson we learn from history is that we never learn from history.'

My cardinal's bigger than yours

Over the last twenty years or so, logicians have studied all sorts of alternatives to standard set theory, using the axiomatic approach that goes back to Euclid. They have discovered all sorts of straightforward-looking theorems whose truth or falsity depends on the precise choice of axioms. And they have managed to organize a lot of this material in terms of axioms for the existence of 'large cardinals'. For example an *inaccessible* cardinal is one not expressible in terms of a smaller number of smaller cardinals. Roughly, the type of cardinal possible in a set theory pretty much determines everything else it can do.

Mathematical logicians study the relative consistency of different axiomatic theories in terms of *consistency strength*. One theory has greater consistency strength than another if its consistency implies the consistency of the other (and, in particular, if it can model the other). The central problem in mathematical logic is to determine the consistency strength in this ordering of any given piece of mathematics. One of the weakest systems is ordinary arithmetic, as formalized axiomatically by Giuseppe Peano. But even weaker systems have been found useful. Analysis finds its place in a stronger theory, called second-order arithmetic. Still stronger theories arise when we axiomatize set theory itself. The standard version, *Zermelo-Frankel Set Theory*, is still quite weak, although the gap between it and analysis is large, in the sense that many mathematical results require more than ordinary analysis, but less than the whole of Zermelo-Frankel Set Theory, for their proofs. Beyond Zermelo-Frankel, large cardinals hold sway. For example R. M. Solovay showed that the axiom 'there exists an inaccessible cardinal' implies that every set of reals is Lebesgue measurable (see chapter 14); and subsequently Saharon Shelah proved the converse. There are other even larger types of cardinals: Mahlo, weakly compact, hyper-Mahlo, ineffable, measurable, Ramsey, supercompact, huge, n-huge. It's enough to make Kronecker turn in his grave—infinitely many times.

8

Ghosts of departed quantities

Mad Mathesis alone was unconfined,
Too mad for mere material chains to bind,
Now to pure space lifts her ecstatic stare,
Now, running round the circle, finds it square.

Alexander Pope

In these profligate times it is hard to realize what a precious commodity paper used to be. Richard Westfall tells us that in 1612 the rector of North Witham, one Barnabas Smith, 'entered a grandly conceived set of theological headings' in a huge notebook, 'and under these headings a few pertinent passages culled from his reading. If these notes represent the sum total of his lifetime assault on his library, it is not surprising that he left no reputation for learning. Such an expanse of blank paper was not to be discarded in the seventeenth century.' I mention all this for one reason only: the treasure was seized by Smith's stepson, Isaac Newton, who called it his 'Waste Book', and it records the birth pangs of the calculus and mechanics.

In medieval times paper (rather, parchment) was even scarcer, and old manuscripts were habitually washed clean by monks seeking a surface on which to inscribe matters of high religious import. Imperfect erasure would leave faint traces of the original, known as a *palimpsest*. Where the medieval monks washed well, they destroyed works of enormous value; but more than one important document has been inadvertently saved for posterity by bad laundering. In 1906 the Danish scholar J. L. Heiberg heard of a mathematical palimpsest at Constantinople, tracked it down, and realized that it had originally contained works of Archimedes. These had been effaced in order to make room for a thirteenth-century Euchologion—a collection of prayers and liturgies of the

Eastern Orthodox Church. The text included *On the Sphere and Cylinder*, much of *On Spirals*, parts of the *Measurement of a Circle, Equilibrium of Planes*, and *Floating Bodies*. All of these works had been preserved in other manuscripts. But, sensationally, the palimpsest also contained the first copy ever found of a long-lost work, *The Method*. It was known from references elsewhere that Archimedes had produced a book with this title, but the contents were a mystery.

It has mathematical as well as historical significance, because *The Method* lets us in on how Archimedes thought of his ideas. Like most mathematicians, he first obtained his results by totally unrigorous methods, and then polished them up into a decent proof. And of course, only the polished version was ever displayed for public consumption. It's a habit that has deprived generations of researchers of the insights of their predecessors. Among the heuristic methods Archimedes used was a technique of slicing solids into infinitely many pieces of infinitesimal thickness, and hanging these on the arms of a notional balance, where their sum could be compared with some known object. He found the volume of a sphere this way. A great many problems are easier to solve rigorously if you know in advance what the answer is, so the method was of great value to Archimedes. Other mathematicians have made similar use of 'infinitesimal' arguments, and usually (but not always) apologized for their lack of rigour, waved their hands a bit, and asserted that 'of course it can all be made rigorous if you take a bit more trouble'.

Infinitesimals plagued seventeenth- and eighteenth-century analysis, although the philosophical arguments about them didn't prevent a lot of good work being done. By the nineteenth century the notion of an infinitesimal had been safely banished and replaced by that of a limit. It was generally agreed that infinitesimals were nonsense. Which only goes to show that you can't always trust a consensus view . . .

Objections from Elea

As we have seen, the Pythagorean ideal of number as the basis for everything fell foul of the irrational. It also ran into philosophical trouble with the Eleatic school, the followers of Parmenides of Elea. The best known of these is Zeno, who

lived around 450 BC. Zeno proposed four paradoxes about
the nature of space and time. Two attack the idea that they
are discrete; two that they are continuous. The story of the
infinitesimal is much less straightforward than that of its cousin
the infinite, and considerations such as Zeno's played a significant role. The 'dichotomy' paradox attacks infinite divisibility of space. Before an object can travel a given distance, it
must first travel half as far. But before doing that, it must
travel a quarter as far, and so on. Trying to do infinitely many
things in the wrong order, it can never get started. 'Achilles
and the tortoise' involves a similar scenario, with a speedy but
late-starting Achilles unable to catch a more tardy tortoise
because whenever Achilles reaches the place where the tortoise *was*, it has moved on. The 'arrow' attacks discreteness.
An arrow in flight is in a fixed position at a fixed instant—but
is then indistinguishable from a motionless arrow in the same
position, so how do we know it's moving? The 'stade' is more
involved, and attacks the idea that both space and time can
only be subdivided a definite amount. Zeno's paradoxes are
more subtle than they seem, and as assertions about physical

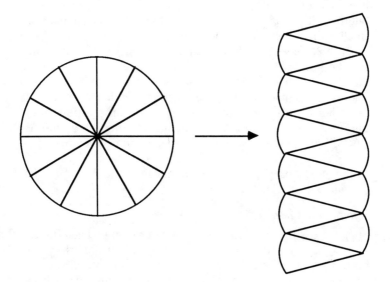

Fig. 11.   Dissecting a circle into a near-rectangle.

space-time, rather than mathematical idealizations, still pose tricky questions. The Greeks found the paradoxes devastating, and this pushed them even further away from numbers and into the arms of geometry.

Venerable beads

Can a line be thought of as some kind of sequence of points? Can a plane be sliced up into parallel lines, or a solid into planes? The modern view is 'yes', the verdict of history an overwhelming 'no'; the main reason being that the interpretation of the question has changed.

If a line can only be subdivided a definite amount, then points are like little 'atoms of lineness', and follow each other like beads on a string. After each point there is a unique *next* point. So what real number corresponds to the point 'next to' the origin? Is it 0.001? No, because 0.0001 is closer. But 0.00001 is closer still, and 0.000001 closer than that, and . . . one starts to feel sympathy for Achilles. What we want to do is write down 0.00000 . . . and then put a 1 in the very last place, but there *is* no last place! There are two ways out of the dilemma. One is to assert that there really is a next number, but that it is *infinitesimally* larger than 0, which means it's smaller than anything of the form 0.00 . . . 01. The other is to accept that there is no 'next number' larger than 0. Then a line can be subdivided indefinitely, and there are no 'ultimate atoms' or *indivisibles*. Which means that you can't think of a line as being made up of points strung together in order. On the other hand it's obvious that any particular position on the line *is* a point: just draw another line crossing there, and use Euclid's axiom that 'two lines meet in a point'. It's a real bind, isn't it?

Our next observation about infinitesimals looks as if it has the game pretty well sewn up:

Infinitesimals don't exist

Suppose x is the smallest number greater than 0. Then $0 < x/2 < x$ so $x/2$ is also greater than zero, and smaller than x to boot. Contradiction.

However, we shouldn't be too easily satisfied, because another observation is:

Infinitesimals are useful

I've already mentioned Archimedes weighing infinitesimals to find the volume of a sphere. Nicholas of Cusa found a similar approach to the area of a circle, by slicing it up like a pie. Imagine a large number of very thin slices. Each is almost triangular, so its area is half the product of its height (the radius r of the circle) and the length of the base. The total length of the bases is near enough the circumference of the circle, which we know is $2\pi r$. So the area is roughly 1/2 times r times $2\pi r$, that is, πr^2. If we could think of the circle as an infinite number of infinitesimally thin triangles, there would be no error at all. This is such a simple proof when compared to a Eudoxan style exhaustion that it's a pity it's nonsense. On the other hand, if it's nonsense, why does it give the right answer? And it's not the only example: for instance Democritus found the volume of a cone by piling up circular slices, an argument that attracted Archimedes's admiration.

From the time of Apollonius, about 250 BC, there was great interest in the 'problem of tangents': how to draw a tangent to a given curve. Apollonius solved it for conic sections, in particular the parabola, by purely geometric means. Newton and Leibniz, developing the calculus, attacked the problem using coordinate geometry. The equation of a parabola is $y = x^2$. The main problem is to find the *slope* of its tangent at x. Newton argued along the following lines. Let x increase slightly to $x + o$. Then x^2 changes to $(x + o)^2$. The *rate* of change is therefore the ratio of the difference between the squares to the difference in the value of x, namely

$$[(x + o)^2 - x^2]/[(x + o) - x],$$

which simplifies to yield

$$[2ox + o^2]/o = 2x + o.$$

Let o approach zero; then the slope approaches $2x + 0 = 2x$. This is the slope of the tangent, or, as Newton put it, the *fluxion* of the *fluent* x^2. The answer agrees with Apollonius, too. Another coincidence? Newton's method, the wellspring of modern analysis, refreshes parts that Apollonius's cannot reach. It works just as well on $y = x^3$, giving a fluxion $3x^2$.

Leibniz had a similar argument, but instead of o he used the symbol dx ('a little bit of x').

It may strike you that there's something fishy about all this. If so, you're in agreement with the arch-critic of calculus, Bishop George Berkeley. In 1734 he published *The Analyst, Or a Discourse Addressed to an Infidel Mathematician. Wherein It is examined whether the Object, Principles, and Inferences of the modern Analysis are more distinctly conceived, or more evidently deduced, than Religious Mysteries and Points of Faith. 'First cast out the beam in thine own Eye; and then shalt thou see clearly to cast out the mote out of thy brother's Eye.'* Berkeley objected that either o is not exactly zero—in which case the answers are wrong, albeit not by much—or it *is* zero, in which case you can't divide by it, and the calculation doesn't make sense. He suggested that the use of little o rather than big 0, while not necessarily exactly *cheating*, doesn't really improve matters.

Newton had defined a fluxion as the 'ultimate ratio of evanescent increments'. To this Berkeley reasonably responded: 'And what are these fluxions? The velocities of evanescent increments. And what are these same evanescent increments? They are neither finite quantities, nor quantities infinitely small, nor yet nothing. May we not call them ghosts of departed quantities?' Berkeley was objecting to the calculation of a fluxion as a *ratio* between two quantities ($2ox + o^2$ and o, in the example above) which both vanish. The ratio 0/0 makes no sense. Since $1 \times 0 = 0$ we have $0/0 = 1$; but $2 \times 0 = 0$ so $0/0 = 2$. In fact you can similarly 'prove' that $0/0 = x$ for any x you please. So why does Newton's method seem to work? According to Berkeley, because of compensating errors.

The first thing to realize about Berkeley's objections is that he had the analysts dead to rights. It was a penetrating, informed criticism, and it revealed a horrible gap. He may have been grinding an axe about the difference (or lack thereof) between mathematical proof and religious belief, but he'd thought very hard about what he was criticizing. (Would that others might do the same.) Nobody had a *really* good answer to Berkeley. On the other hand it just won't wash to say that hundreds and hundreds of calculations, all confirmed in other ways, are due to 'compensating errors', without explaining why the devil the errors are perpetually conspiring to compensate! Indeed, if you can *prove* that the errors always compensate, that is tantamount to proving they aren't errors at all. Moreover, Newton and his followers knew that what they were doing made sense, because they had a slightly more subtle picture (albeit physical) of what was happening. Berkeley insisted on thinking like an algebraist, with *o* being some definite constant. Then *o* must be both zero and non-zero, an absurdity. But Newton thought of *o* as a *variable*, which can be made to *approach* 0 as closely as we please, without actually vanishing. Newton thought of the whole *process* whereby *o* shrinks to zero; Berkeley wanted Newton to fix its value at the instant before it disappeared forever, which he couldn't do. Newton knew that Achilles will catch the tortoise eventually because he is running faster; Berkeley was trying to make Newton isolate the instant immediately before the tortoise is overtaken, it being clear that there is no such thing. Berkeley won the logical argument hands down. Nobody took the slightest notice, because the method worked beautifully, logic or no.

Near misses

Colin MacLaurin, in his *Treatise on Fluxions*, tried to prove the correctness of calculus by reinterpreting its methods within classical geometry, notably exhaustion. He became such a dab hand at this that he persuaded several people to abandon calculus altogether in favour of geometry, which wasn't his intent. Leibniz sought refuge in the infinitesimal. 'Now these d*x* and d*y* are taken to be infinitely small. . . . It is useful to consider quantities infinitely small such that when their ratio is sought, they may not be considered zero but which are rejected as often as they occur with quantities incomparably greater.' He also thought of an infinitesimal as a variable quantity; and he summed it all up in a *Law of Continuity*: 'In any supposed transition, ending in any terminus, it is permissible to institute a general reasoning, in which the final terminus may also be included.' In other words, what works for all non-zero *o*s also works for *o* = 0. He supported this principle by examples, not proof. It needs caution in use: try it on the 'general reasoning' *o* > 0. John Bernoulli wrote part

II of a calculus text in 1692, eventually published in 1742. Publication delays in mathematics are not unknown, but ones of this magnitude are unusual. Perhaps Bernoulli was trying to set a record: the first part, written in 1691, wasn't published till 1924. Seeking to clarify the mystery, he explained that 'a quantity which is diminished or increased by an infinitely small quantity is neither increased nor decreased'. Is it only the modern eye that finds this enigmatic? He also tried to define the infinitesimal as $1/\infty$.

Newton's ideas of 1704 got closer than Bernoulli's. 'In mathematics the minutest errors are not to be neglected. I consider mathematical quantities not as consisting of very small parts, but as described by a continual motion. Fluxions are, as near as we please, as the increments of fluents generated at times, equal and as small as possible.' He explicitly rejected the infinitesimal justification as being incoherent. He got very close indeed to the modern interpretation when, in the first and third editions of the *Principia*, he said: 'Ultimate ratios in which quantities vanish are not, strictly speaking, ratios of ultimate quantities, but limits to which the ratios of these quantities, decreasing without limit, approach, and which, though they can come nearer than any given difference whatever, they can neither pass over nor attain before the quantities have diminished indefinitely.' In other words, to find the limit which the ratio $2ox + o^2 : o$ approaches as o diminishes towards 0, you do *not* set $o = 0$ to get the ratio $0:0$. Instead you keep o non-zero, simplify to get $2x + o : 1$, and then observe that as o *approaches* 0, this ratio *approaches* $2x:1$. Newton knew exactly what he was doing; he just couldn't find a precise way to express it. From about this time a series of mathematicians, including John Wallis and James Gregory, missed the target by a hair's breadth.

The plot thickens

Before calculus had been properly sorted out, a new complication arose: complex numbers. As we shall see in chapter 11, the extension of analysis to complex-valued functions was itself a source of great (but creative) confusion. The foundational problems of calculus were soon to be settled, though. In the bible of complex analysis, his *Cours d'Analyse* of 1821,

Augustin-Louis Cauchy based the theory on the idea of a *limit*, and defined it thus: 'When the successive values attributed to a variable approach indefinitely a fixed value so as to end by differing from it by as little as one wishes, this last is called the limit of all others.' As regards the infinitesimal, he was quite explicit: 'One says that a variable quantity becomes infinitely small when its numerical value decreases indefinitely in such a way as to converge to the limit 0.' Infinitesimals are *variables*, not constants. Similarly ∞ is not a constant, but a variable that becomes indefinitely large. What we have here is the Aristotelian notion of *potential* infinity, transmuted into potential infinitesimality. It is perhaps churlish to remark that nobody had defined what they meant by a variable. The omission was not serious, and went unnoticed.

There is a Larry Niven story, *Convergent Series*, in which a bright red, horned demon is raised using the traditional method, a pentacle, in which it appears spreadeagled. The protagonist is granted one wish, but valid for a mere twenty-four hours because of the usual small print in the contract. After that the demon reports to Hell, instantly reappears inside the pentacle, and exacts the time-honoured penalty. The victim does have the option of erasing the pentacle and redrawing it wherever he pleases. Niven's hero scrawls his pentacle on the demon's bulging belly. When the demon reappears if finds itself shrinking rapidly, and before it can work out what's happening . . . Are we left with an infinitesimal demon? No, says Cauchy: we are left with no demon at all. It is the *way it vanishes* that is infinitesimal.

Epsilontics

Finally—and frankly it's a relief to see it—Karl Weierstrass sorted out the muddle in 1850 or thereabouts by taking the phrase 'as near as we please' seriously. How near *do* we please? He treated a variable, not as a quantity actively changing, but simply as a static symbol for any member of a *set* of possible values. A function $f(x)$ approaches a limit L as x approaches a value a if, given any positive number ε, the difference $f(x) - L$ is less than ε whenever $x - a$ is less than some number δ *depending on* ε. It's like a game: 'You tell me how close you want $f(x)$ to be to L; then I'll tell you how close

x has to be to a.' Player Epsilon says how near *he* pleases; then Delta is free to seek his own pleasure. If Delta always has a winning strategy, then $f(x)$ tends to the limit L. This epsilon–delta definition of limit is perhaps a little cumbersome, but, like the Greek method of exhaustion, a competent professional soon gets accustomed to it, and can wield it with precision and occasionally dazzling virtuosity.

Notice how the physical ideas of motion have been replaced by a set of static events, one for each choice of ε. It is not necessary to think of a variable ε flowing towards 0; all we have to do is entertain all possible values (greater than 0) for ε, and deal successfully with each. The introduction of potential versus actual infinitesimality is a red herring; the whole problem can be formulated in purely finite terms. Weierstrass's definition of limit freed calculus from metaphysical considerations and modern analysis was born. Today every undergraduate beginning an analysis course is propelled through a week or two of what is colloquially called 'epsilontics'—usually to his or her initial perplexity. It's the way the trade bloods its apprentices. If you want to *understand* analysis, and not merely calculate with calculus, this is the approved path.

Reeb's dream

For all that, there's something intuitive and appealing about the old-style arguments with infinitesimals. They're still embedded in our language, in the way we think: 'instants' or 'moments' of time; 'instantaneous' velocities; the idea of a curve as a series of infinitely small straight lines. Engineers, chemists, and physicists happily introduce small increments δx, which miraculously transmogrify into dx at judicious points in the calculation, without worrying about full-blooded epsilontics. 'The mathematicians have justified this kind of thing anyway, let them beat their tiny brains out on it if they want to, I'm after bigger game.'

Georges Reeb puts it eloquently:

The dream of an infinitesimal calculus worthy of the name, that is to say in which dx and dy are infinitesimal numbers, $\int_b^a f(x)\,dx$ is a genuine sum of such numbers, and limits are attained, has always been dreamed by mathematicians; and such a dream deserves perhaps an epistemological enquiry. Some other dreams, lesser maybe if compared with the achievements of calculus, have haunted the mathematician's world and wishful thought: it is the idea of a world where integers can be classified as 'large', 'small', or even 'indeterminate' without loss of consistent reasoning, satisfy the induction principle, and where the successors of small integers would remain small; a world where concrete collections, fuzzy perhaps but anyhow not finite, could be gathered in a single finite set; a world where continuous functions would be approximated almost perfectly by polynomials of a fixed degree. In such a world, the finite realms could be explored either through the telescope or through the magnifying glass in order to gather entirely new pictures. Within such a world, the criteria of rigour set forth by Weierstrass, interpreted in a two-fold sense, would allow for phantasy and metaphor.

A footnote adds an example: 'There should be a finite chain linking some monkey to Darwin, respecting the rules: a monkey's son is a monkey, the father of a man is a man.'

Non-standard analysis

Between about 1920 and 1950 there was a great explosion of mathematical logic; and one of the topics that emerged was Model Theory. In model theory one contemplates some system of axioms, and attempts to find and characterize mathematical structures that obey all the axioms of the system— *models* of the axioms. Thus the coordinate plane is a model for the axioms of Euclidean geometry, Poincaré's universe is a model for the axioms of hyperbolic geometry, and so on.

The logicians found an important distinction between what they called first-order and second-order axiomatic systems. In a first-order theory the axioms can express properties required of all objects in the system, but not of all *sets* of objects. In a second-order theory there is no such restriction. In ordinary arithmetic, a statement such as

$$(1) \quad x + y = y + x \text{ for all } x \text{ and } y$$

is first order, and so are all the usual laws of algebra; but

$$(2) \quad \text{if } x < 1/n \text{ for all integers } n \text{ then } x = 0$$

is second order. The usual list of axioms for the real numbers is second order; and it has long been known that it has a *unique* model, the usual real numbers R. This is satisfyingly tidy. However, it turns out that if the axioms are weakened, to comprise only the first-order properties of R, then other models exist, including some that violate (2) above. Let R* be such a model. The upshot is a theory of *non-standard analysis*, initiated by Abraham Robinson in about 1961. In non-standard analysis there are actual infinities, actual infinitesimals. They are constants, not Cauchy-style variables. And Reeb's Dream—comes true! The violation of (2), for example, means precisely that x is a 'genuine' infinitesimal. And if ω is an infinite integer, then the sequence

$$1, 2, 3, \ldots, \omega - 3, \omega - 2, \omega - 1, \omega$$

connects the monkeys 1, 2, 3, . . . to the Darwins . . . , $\omega - 3$, $\omega - 2$, $\omega - 1$, ω.

Infinitesimals do exist

It's not even a new game. How many times before has the system of 'numbers' been extended to acquire a desirable property? From rationals to reals to allow $\sqrt{2}$; from reals to complexes to allow $\sqrt{-1}$. So why not from reals to hyperreals to allow infinitesimals? In non-standard analysis there are ordinary natural numbers N = {0, 1, 2, 3, . . . } all right; but there is a larger system of 'unnatural', better 'non-standard natural' numbers N*. There are the ordinary integers Z, and the non-standard integers Z*. And there are the standard reals R, plus non-standard reals R*. And each pair is indistinguishable as far as first-order properties go, so you can prove first-order properties of R by working with R* if you want; but R* has all sorts of new goodies, like infinitesimals and infinities, which you can exploit in new ways.

First, a few definitions to give the flavour. A hyperreal number is *finite* if it is smaller than some standard real. It is *infinitesimal* if it is smaller than all positive standard reals. Anything not finite is *infinite*, and anything not in R is *non-standard*. If x is infinitesimal then $1/x$ is infinite, and vice versa. Now for the clever bit. Every finite hyperreal x has a unique *standard part* std(x) which is infinitely close to x, that is, $x - $ std(x) is infinitesimal. Each finite hyperreal has a unique expression as 'standard real plus infinitesimal'. It's as if each standard real is surrounded by a fuzzy cloud of infinitely close hyperreals, its *halo*. And each such halo surrounds a single real, its *shadow*.

Let's take a look at Newton's calculation of the fluxion (derivative) of the function $y = f(x) = x^2$. What he does is take a small number o, and form the ratio $[f(x + o) - f(x)]/o$. This boils down to $2x + o$, which would be great if the o could be got rid of. Hence ultimate ratios, limits, and epsilontics, as we saw. But there's a simpler way. Take o to be infinitesimal, and instead of $2x + o$ take its *standard part*, which is $2x$. In other words, the derivative of $f(x)$ is defined to be

$$\text{std}\{[f(x + o) - f(x)]/o\}$$

where x is a standard real and o is any infinitesimal. The Alexandrine sword that slices the Gordian Knot is the innocent-looking idea of the standard part. It's perfectly rigorous, because $\text{std}(x)$ is uniquely defined. And it doesn't just *forget* about the extra o; it really does get rid of it altogether. What's going on here? Basically, there are orders of magnitude for hyperreals, and you must work to the right order. Standard reals are order 0. Infinitesimals like o are mostly order 1, but o^2 is order 2, o^3 order 3, and so on. Meanwhile infinities such as $1/o$ are order -1, $1/o^2$ is order -2. The standard part picks out the lowest-order term in any finite hyperreal.

Non-standard analysis does not, in principle, lead to conclusions about R that differ in any way from standard analysis. It is the method, and the setting in which it operates, that are non-standard; but the results are true theorems of good old-fashioned analysis. So what's the point of all this rigmarole, if it can't prove anything new? Before discussing this important question, let's get some kind of feel for how non-standard analysis behaves.

Welcome to the madhouse

A course in non-standard analysis looks like an extended parade of exactly those errors that we spend all our time teaching students to avoid. For example:

(1) A sequence s_n is bounded if s_ω is finite for all infinite ω.

(2) A sequence s_n converges to a limit L if $s_\omega - L$ is infinitesimal for all infinite ω.

(3) A function f is continuous at x if $f(x + o)$ is infinitely close to $f(x)$ for all infinitesimal o.

(4) The function f has derivative d at x if and only if $[f(x + o) - f(x)]/o$ infinitely close to d for all infinitesimal o.

(5) The Riemann integral of f is an infinite sum of infinitesimal areas.

However, within the formal framework that non-standard analysis provides, these statements are not erroneous: they are rigorous truths!

Why bother?

Back of my question: what's the point of all this rigmarole, if it can't prove anything new?

The question is an oversimplification. The correct statement is that any theorem proved by non-standard methods is a true theorem of standard analysis (and therefore must have a standard proof). But that doesn't tell us how to find the standard proof, nor does it give any idea whether the non-standard proof will be shorter or longer, more natural or more contrived, easier to understand or harder. As Newton showed in his *Principia*, anything that can be proved with calculus can also be proved by classical geometry. In no way does this imply that calculus is worthless. In the same way, the correct question is whether non-standard analysis is a more powerful piece of machinery. That's not a matter for mathematical argument: it has to be resolved by experience. Experience suggests quite strongly that proofs via non-standard analysis tend to be shorter and more direct. They largely avoid the use of complicated estimates of the sizes of things. Instead of worrying whether o^2 is less than $10^{-93.7}$ or whatever, we merely observe that it's a second-order infinitesimal, and so on. Such estimates form the bulk of the proofs of classical analysis. The main difficulty is that non-standard analysis requires a very different background from classical analysis, so a lot of initial effort must be put in before you come to the pay-off.

The canard unchained

New phenomena have been discovered by non-standard analysis, including some of importance in applications. One vast area of applied mathematics, perturbation theory, deals with 'very small' changes to equations. Why not make the perturbation infinitesimal, and see what you get? A few brave souls have embarked on this course in recent years. One example must suffice. There is a certain differential equation, which I won't write down, that involves a small parameter ε and an additional parameter k. It is known that for $k < 1$ there is a unique oscillatory solution, whereas for $k \geq 1$ no oscillations exist. How does the oscillation 'switch on', as k passes through 1? There is a mathematical gadget called a 'Hopf bifurcation' that acts as a prototype for the onset of an oscillation. The oscillation just grows smoothly from a small one to a larger one. It's easy to calculate that a Hopf bifurcation occurs in the equation at $k = 1$. What more need be said?

Apparently, a great deal. As k decreases through 1 a small oscillation does indeed switch on by a Hopf bifurcation. But almost immediately, it explodes *very* rapidly into a big one. Because some of the intermediate pictures involve curves shaped like ducks, the French discoverers of this effect called it a *canard*. You need an accurate computer to bag a duck: for example when ε = 0.01 it is necessary to look at the sixtieth decimal digit of k. The first duck was found by Eric Benoit, Jean-Louis Callot, and Francine and Marc Diener. And they used no computer, but non-standard analysis, with an infinitesimal perturbation ε. The duck promptly surrendered of its own accord, overwhelmed by the power of the method.

The infinitesimal computer

One of the big problems in computer graphics is that the plane is continuous but the computer screen is discrete: a rectangular array of pixels, tiny squares that are coloured black or white (or red or green or other exotic colours) to represent shapes. Traditional geometric objects, such as circles and lines, live in the continuous plane; and so does the mathematical theory that describes them. The discrete lattice structure of the computer screen causes all sorts of difficulties: traditional continuous geometry is the wrong tool, but what else is there?

It is, for example, surprisingly hard to decide whether or not a given collection of black pixels represents a straight line. On a discrete lattice, two straight lines can cross without having any pixels in common; or—if nearly parallel—they can meet in what appears to be a long segment. And while translating objects is relatively easy (at least through distances represented on the lattice), rotating them is somewhat fraught. If you rotate a square pixel through 45° it ends up diamond-shaped and overlaps *several* pixels. It's hard to handle points that start to multiply when you rotate them.

The French mathematician Jean-Pierre Reveillès has enlisted the aid of non-standard analysis to overcome this difficulty. The basic idea is to model the discrete but finite computer screen as a discrete but *infinite* lattice of infinitesimal pixels. From the philosophical viewpoint of non-standard analysis this is a finite object. It requires just one infinite integer, call it ω, and it provides an accurate model of the continuous plane of standard analysis. Computer programs for doing things in plane geometry can be rewritten using expressions in ω instead of real numbers, leading to a 'computer language' that is quite simple to use.

All very well, but of course a real computer doesn't have an infinite screen with infinitesimal pixels. This is where the clever bit comes in. Take the computer program, approximate the infinite integer ω by a moderately large finite one, say 10,000. The result is a program that runs on a real computer, using ordinary arithmetic, and which computes the desired geometric object to a high degree of accuracy. If you want more accuracy, just replace 10,000 by 1,000,000 or something larger; if you want a quick-and-dirty computation, reduce it to 1,000 or 100.

This approach is a remarkably 'clean' way to formalize discrete graphics. It handles difficulties such as those mentioned above with ease; it deals with rotations on the same basis as translations. As a consequence of its clean structure, programs written in this manner tend to be unusually efficient, often running in a fraction of the time taken by more conventional programs. Not only that: the method opens up new lines of attack in 'standard' combinatorics, and answers long-standing questions—for instance, providing formulas for the number of lattice points inside a given triangle.

It's unlikely that an approach of this kind would ever have emerged from conventional computer science: it transports the problem into highly abstract regions, returning to the concrete world only at the very end of the analysis. Remarkably, it provides not only a practical method for making discrete graphics work, but arguably one of the best available. Non-standard analysis comes into its own when it does *not* try to mimic standard analysis, but is given its head and allowed to follow its own individual track.

Logic for engineers?

Indeed there are reasons to expect that it is in applied mathematics that the most useful products of non-standard analysis might emerge. First, because applied mathematicians, engineers, and scientists already think in terms of infinitesimals like dx. Second, because so much attention is paid to small perturbations. And too much of perturbation theory is hopelessly unrigorous and disorganized, and reeks of rule-of-thumb. There's a good chance that non-standard analysis could bring some much-needed order into this area, and add some power to its elbow. The non-standard analysts have explored some of the possibilities, including polynomial equations, fast-slow flows, geodesics, boundary value problems, boundary layer flow, stochastic differential equations, and spin systems in mathematical physics.

In scientific research, one might expect new ideas from elsewhere, new techniques, new phenomena, to be seized upon and exploited. Actually this only happens if they don't have to cross traditional boundaries between subjects. The 'Not Invented Here' syndrome is very strong, and some perturbation theorists appear to be suffering from it. For example, the reaction of one of them to the discovery of canards was to find a (rather complicated and delicate) proof from his own point of view, while expressing doubts (unjustified) as to the rigour of the non-standard approach. The pot calling the kettle black, in spades! It's easy to discover things with hindsight: I'd be more impressed by the perturbation theory proof if it had been found *before* the non-standard analysts had done the donkey-work.

But the reaction is symptomatic of a genuine problem, by no means confined to applied mathematics. The people most knowledgeable about the area of application are not, by training, equipped to handle a technique that developed out of mathematical logic—especially one which requires a distinctly different cast of mind, a new style of thinking that takes several years to get used to. They aren't impressed by the way it makes sloppy thinking rigorous, because—dare I say it—they'd rather stay sloppy. And so it will be the non-standard analysts who have to push ahead with the applied development, until they finally come up with something *so* dramatic that even the least imaginative perturbation-theorist will begin to feel that maybe he's missing out on something. Of course, it may never happen—but I rather suspect it will, for the times they are a-changing. However, whether the engineers of the twenty-first century will have to study the metamathematics of model theory remains open to doubt.

9

The duellist and the monster

At the end of a session at the Engineering
College, Cooper's Hill, a reception was held and
the science departments were on view. A young
lady, entering the physical laboratory and seeing
an inverted image of herself in a large concave
mirror, naively remarked to her companion:
'They have hung that looking glass upside
down.'

Clement V. Durell

'Pistols at twenty-five paces!' On 30 May 1832 a young French
revolutionary, duelling with a comrade over the honour of a
lady, was hit in the stomach with a pistol-ball and died of
peritonitis a day later. He was buried in the common ditch at
the cemetery of Montparnasse. Just before the duel he wrote
to his friends Napoleon Lebon and V. Delauney: 'I die the
victim of an infamous coquette. It is in a miserable brawl that
my life is extinguished. Oh! Why die for so trivial a thing, for
something so despicable!' The circumstances are mysterious;
indeed until recently the lady was known only as 'Stéphanie
D.' until Carlos Infantozzi did some detective work and revealed her as Stéphanie-Felicie Poterin du Motel, the entirely
respectable daughter of a physician.

The protagonist in the melodrama wrote another letter that
night, to Auguste Chevalier, saying, 'I have made some new
discoveries in analysis.' And he ended, 'Ask Jacobi or Gauss
publicly to give their opinion, not as to the truth, but as to the
importance of these theorems. Later there will be, I hope,
some people who will find it to their advantage to decipher
all this mess.' The mess was, according to Tony Rothman,
'instrumental in creating a branch of mathematics that now
provides insights into such diverse areas as arithmetic, crystallography, particle physics, and the attainable positions of

Rubik's cube'. The tragic creator was Evariste Galois; the branch of mathematics was Group Theory.

What shape is an equation?

Galois was working on a long-standing problem. As we saw in chapter 1, equations of degrees 1, 2, 3 and 4 can be solved by an exact formula, an algebraic expression involving nothing worse then nth roots, or *radicals*. But the equation of degree 5, the *quintic*, resisted all efforts to find such a formula. By the 1800s attention had switched, not to finding a solution, but to proving that no solution by radicals exists. Paolo Ruffini made an abortive attempt in 1813, and the question was finally settled in 1824 by Abel. Abel's proof involved a complicated analysis of the degrees of auxiliary equations that might arise during the solution process, assuming there was one. Galois was more ambitious. Even though the general equation of degree 5 is insoluble, various special equations of equal or higher degree do possess solutions by radicals. Galois wanted a method to decide, for *any* given equation, whether or not in can be solved by radicals. And his answer was striking and original. Everything depends on the *symmetries* of the equation.

What is symmetry?

'Symmetry', said Hermann Weyl in a famous series of lectures on the topic, 'is one idea by which man through the ages has tried to comprehend and create order, beauty, and perfection.' When mathematicians speak of symmetry they have something very specific in mind: a way of transforming an object so that it retains its structure.

For example, consider a square, and draw its outline around it. If you now move the square, it doesn't usually fit back into the outline. But for exactly eight different motions, it does. These motions are:

(1) Leave it alone.
(2) Rotate 90° (anticlockwise).
(3) Rotate 180°.
(4) Rotate 270°.
(5) Flip it over about a vertical axis (like a revolving door).

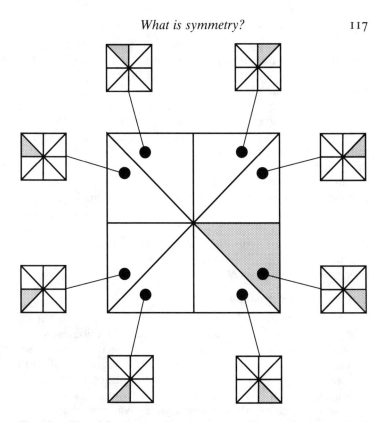

Fig. 12. The eight symmetries of a square. The small squares show where the shaded triangle moves under each of the eight symmetry operations.

(6) Flip it over about a horizontal axis (like a cat-flap).
(7) Flip it about one diagonal.
(8) Flip it about the other diagonal.

Each of these operations is a *symmetry* of the square. If two symmetries are applied in turn, then the square obviously still ends up inside its outline; so the result is another symmetry. For example, 'rotate 90°' followed by 'rotate 180°' produces the same effect as 'rotate 270°'. We say that the set of symmetries has the *group property*, meaning that the 'product' of any two symmetries is a symmetry. The set of symmetries is called the *symmetry group* of the square. This particular group contains eight symmetry operations: we say it has *order* eight.

Consider now an infinite floor, covered with square tiles in the usual fashion. This tiling pattern has all of the eight symmetries of the square; but in addition it has infinitely many *translational* symmetries. The whole pattern can be displaced 1, 2, 3, . . . units horizontally or vertically, up or down, left or right, as well as being rotated or flipped. So there are infinitely many symmetries, and the symmetry group has infinite order.

Any mathematical object, however abstract, can possess symmetries. These are now defined as transformations of the object that preserve its 'shape', that is, its basic structure. The symmetries again form a group; that is, if two are performed in turn the result is yet another symmetry. The way to study symmetry as a general concept is to look at the algebraic properties of this process of composition of transformations.

The shape of an equation

All this is much clearer now than it was in Galois's day; but I'll take advantage of the modern viewpoint to describe his ideas. So what shape is an equation? The mathematical object involved is the full set of its solutions. (An equation of the nth degree always has n real or complex solutions.) The transformations are the different ways to rearrange these among themselves, that is, their *permutations*. And the shape—the structure to be preserved—is the system of algebraic relations that hold between them. The idea is most easily grasped using an equation whose solutions are easy to find:

$$x^4 - 5x^2 + 6 = 0.$$

The roots are $\sqrt{2}$, $-\sqrt{2}$, $\sqrt{3}$, and $-\sqrt{3}$: call them α, β, γ, δ. Clearly α and β form a 'matching pair', as do γ and δ. In fact there is no way to distinguish α and β from each other, just by using equations with rational coefficients. Thus α satisfies $\alpha^2 - 2 = 0$ but so does β. And $\alpha + \beta = 0$; but interchanging α and β gives $\beta + \alpha = 0$, also true. Similarly γ and δ are algebraically indistinguishable (by equations over the rationals). On the other hand it is easy to tell α from γ, because $\alpha^2 - 2 = 0$ is true, but $\gamma^2 - 2 = 0$ is false. I don't mean that there is no difference at all between $\sqrt{2}$ and $-\sqrt{2}$:

obviously they have different signs. But this difference of sign can't be detected using only polynomial equations with rational coefficients. If we write a list of the four roots, there are exactly four ways to rearrange them so that all true equations in the original order remain true in the new order. They are:

$$\alpha\beta\gamma\delta \mid \alpha\beta\delta\gamma \mid \beta\alpha\gamma\delta \mid \beta\alpha\delta\gamma.$$

This list of permutations is the *Galois group* of the equation. And you can tell, just by looking at the Galois group, whether the equation is soluble by radicals! It stares you in the face, doesn't it? No? Well, it does, if you are a budding Galois: the recipe *is* a little complicated to state. I'll do my best: the main point to grasp is that there is a definite computable criterion in terms of the internal algebraic structure of the Galois group.

Suppose the group is *G*. It may contain within it a smaller system of permutations *H*, which also has the group property. If so, call *H* a *subgroup*. A further technical condition on *H*, called *normality*, also comes into play: let's just agree that this means that *H* sits inside *G* in a particularly nice way. Now Galois proves that an equation is soluble by radicals precisely when:

(1) Its group *G* contains a normal subgroup *H* of smaller order;
(2) The order of *G* divided by that of *H* is prime;
(3) *H* contains a normal subgroup *K* of smaller order;
(4) The order of *H* divided by that of *K* is prime;
(5) And so on until we reach a subgroup containing only one element of *G*.

Such a group is said to be *soluble*. (The idea is that an *n*th root can be obtained from a series of *p*th roots, for various primes *p*. For example the sixth root is the square root of the cube root, since $6 = 2 \times 3$; and 2 and 3 are prime. Each *p*th root corresponds to one step in the chain of subgroups *G*, *H*, *K*, . . .)

The quintic

Obviously you can only use Galois's result if you have a well-developed method for analysing groups, subgroups, normal subgroups, orders of subgroups, and so on. Such methods now exist, inspired in part by Galois's work. One by-product is a satisfying explanation of what is so special about the quintic.

The *general n*th degree equation has no special relations between its roots, so its Galois group is the group of *all* permutations of its *n* roots. This is a very specific group and it's not too hard to find out a lot about it. In particular it can be shown that when $n = 2, 3$, or 4 it is a soluble group, so the equations of those degrees are soluble. In fact, the *methods* used to solve them can be interpreted and organized using group theory. However, when $n = 5$ the group G of all permutations is *not* soluble. G has order 120, and there is a normal subgroup H of order 60. The ratio $120/60 = 2$ is prime; so you can get started all right. But, as Galois proved, the only normal subgroups of H have orders 60 and 1. The only possibility for K is the group of order 1; but then the ratio of orders is $60/1 = 60$, which isn't prime. The group isn't soluble, and neither is the equation.

Genius is its own reward?

Galois worked out the basics of this by the time he was seventeen, and submitted some of his results to the Paris Academy of Sciences. The paper was rejected, the manuscript lost. He submitted his results twice more, and on each occasion the paper was lost. Galois blamed the politically oppressive Bourbon regime, and a society that condemned genius in favour of mediocrity. Actually it may well have been his own fault: he tended not to explain his ideas very clearly, and novelty *plus* obscurity is suicidal, especially from a complete unknown. At any rate, Galois took up revolutionary politics, was expelled from the École Normale, arrested, acquitted, arrested again, and jailed for six months in Sainte-Pélagie. A cholera epidemic had him transferred to a hospital, and he was put on parole. Along with his freedom he experienced his first and only love affair, with the all too literal *femme fatale*, Stéphanie. His work might have been lost along with his life, were it not for Liouville, who later began to look through the papers that Galois had left after his death. On 4 July 1843 Liouville addressed the Academy: 'I hope to interest the Academy in announcing that among the papers of Évariste

Galois I have found a solution, as precise as it is profound, of this beautiful problem: whether or not it is possible to solve by radicals . . .'

Groups galore

Galois's work was one of the clues that led to modern Group Theory. Several authors, notably Cauchy and Joseph-Louis Lagrange, worked on groups of permutations (or 'substitutions'). Camille Jordan systematized much of this in his 1870 *Theory of Substitutions and Algebraic Equations*. Meanwhile other objects, similar to permutation groups but not consisting of permutations, began to make their appearance. For example in 1849 Auguste Bravais used symmetries in three-dimensional space to classify the structure of crystals, leading Jordan to consider groups whose elements are not permutations, but linear transformations.

Another major source of inspiration was the theory of *continuous groups* of Sophus Lie. A square has only eight different symmetries, four rotations and four reflections, so its group of symmetries is *finite*, of order 8. But some objects, such as the circle and sphere, have a continuous range of symmetries. A circle can be turned continuously through any angle; so can a sphere, about any axis. So the group of symmetries of the circle or sphere contains a continuum of elements. A rough measure of the complexity of the group is its *dimension*: how many parameters are required to specify its elements. The symmetry group of the circle is 1-dimensional (specify one angle of rotation). That of the sphere is 3-dimensional (select an axis of rotate about; then use the longitude and latitude of the 'pole' of this axis, plus the angle of rotation). Methods from analysis, such as differentiation and integration, can be applied to such groups. Lie developed a far-reaching theory which has become one of the central areas of modern mathematics: Lie groups.

Complex function theory produced yet another type of group, whose elements are *birational* transformations

$$f(z) = (az + b)/(cz + d)$$

of the complex numbers (see chapter 11). These are important in the theory of elliptic functions, which curiously also interested Abel and Galois. Number theory yielded its own store of 'groupish' objects: the integers modulo n under the operation of addition, the non-zero integers modulo a prime p under multiplication, Gauss's classes of quadratic forms . . .

The Erlangen Programme

Another stimulus to the development of group theory was Felix Klein's programme to unify geometry, on which he lectured at Erlangen in 1872. At that time, geometry had split into a horde of related but distinct disciplines: Euclidean and non-Euclidean geometry, inversive and conformal geometry, projective and affine geometry, differential geometry, and the newly emergent topology. There were even geometries with only finitely many points and lines. Klein sought some semblance of order, and found it in the idea of geometry as the invariants of a group of transformations.

For example, consider Euclidean geometry. The basic notion is that of congruent triangles; and two triangles are congruent if they have the same shape and size. In other words, if one can be transformed into the other by a rigid motion of the plane. Now, said Klein: the rigid motions form a group. So we have a group of transformations of the plane; and the properties studied in Euclidean geometry are those, such as lengths and angles, that do not change under the action of this group. Similarly in hyperbolic geometry the group consists of rigid hyperbolic motions; in projective geometry it is the projective transformations; and in topology it is topological transformations. The distinction between geometries is at base group-theoretic. Not only that, said Klein: sometimes you can use the groups to pass between geometries. If two apparently distinct geometries have basically the same group, albeit in disguise, then they are really the same geometry. He cited many examples. For instance, the geometry of the *complex* projective line is basically the same as the real inversive plane, and this in turn is the same as the real hyperbolic plane.

At a stroke, Klein's insight brought clarity and order to what had thitherto been a confused, if exciting, mess. (One major exception should be noted. Riemann's geometry of manifolds deprived Klein of a clean sweep.) It became possible to compare one geometry with another, to say that one was stronger or weaker than another, to use results from one type to prove theorems in another. Classical geometry is perhaps no longer a central area of mathematical *research*, largely because it was investigated so successfully in the nineteenth century; but it remains (as we shall see) central to *mathematics*. And Klein's Erlangen Programme still has an enormous influence. It is a measure of its success that its influence is not always explicitly perceived: the viewpoint has passed into common currency.

Abstract groups

When mathematicians find themselves proving the same theorem fifteen times in fifteen different contexts, they usually decide that a bit of tidying-up is in order. In this case the tidying was begun by Arthur Cayley, who proposed a more general concept: a set of operators such that, if any two are performed in turn, the result is still in the set. This he called an *abstract group*. But Cayley still wanted his operators to operate *on* something. The final step, made explicit in 1902 by Edward Huntington but probably part of the general mathematical 'folklore', was to define a group axiomatically. This is the modern approach. A *group* is a set G, on which is defined a law of composition, or *product*. If g and h belong to G, the product is denoted gh. (The notation is not intended to imply that gh is g multiplied by h in the usual sense.) This product is required to satisfy four properties:

(1) G is *closed*: if g and h are in G, then so is gh.
(2) There is an *identity* 1 in G such that $1g = g = g1$ for all g.
(3) Every g in G has an *inverse* g^{-1} such that $gg^{-1} = 1 = g^{-1}g$.
(4) The *associative law* $g(hk) = (gh)k$ holds.

The permutation groups of Galois, the transformation groups of Bravais and Jordan, Lie's continuous groups, and Cayley's abstract groups, all satisfy these conditions. Condition (1) is Galois's 'group property'; the others express useful basic properties that automatically hold for transformations, but might not for more abstract objects. For example if G is the set of integers, and the 'product' gh is the sum $g + h$, then all four laws hold. The same goes if G is the integers modulo n. At a stroke a whole morass of special results for different realizations of the group concept was reduced to a few simple general propositions. And mathematics acquired a powerful language to describe and exploit ideas of symmetry.

Atoms of symmetry

The analysis of the quintic equation leads to a group H whose only normal subgroups are trivial (order 1) or the whole of H (here order 60). Such a group is said to be *simple*. If you think of a group as a molecule, then the simple groups are its atoms. The key to understanding all groups is to understand the simple groups. Unfortunately, 'simple' here doesn't imply 'easy'.

The first major accomplishment in this area was the classification of simple Lie groups. Lie discovered four 'families' of simple Lie groups, defined in terms of linear transformations of n-dimensional space. The work was completed by Wilhelm Killing in 1888, and improved by Élie Cartan in 1894. The results deserve mention because simple Lie groups show up all over mathematics, and we shall soon encounter them again. First, there are the four families found by Lie. They have a satisfying and beautiful unified theory. But in addition there are five *exceptional* groups. Their dimensions are 14, 52, 78, 133 and 248, and as these numbers may suggest, they conform to no obvious pattern. Their existence has been described as a 'brutal act of Providence', but they are related to all kinds of strange and beautiful 'accidents' in mathematics, and they add some piquancy to what might otherwise be too bland a confection. They appear to be the visible signs of some deeper pattern that is beginning to make sense, but is still imperfectly understood. Simple Lie groups are currently fashionable in mathematical physics, and the exceptional groups appear to have important implications for the possible structures of space-time. An act of Providence they may be; but brutal they are not.

The ten-thousand-page proof

The more elementary problem of classifying the *finite* simple groups proved much less tractable. The additional analytic structure of Lie groups, while making it harder to discuss them without technicalities, provides a lot of mathematical leverage for proving theorems. Information on finite simple groups accumulated slowly during the 1900s, but then nothing much happened until 1957 when the floodgates burst.

The easiest simple groups to find are the 'cyclic' groups formed by the integers modulo a prime, under addition. Not only do they have no normal subgroups: they have no subgroups at all. Next are the alternating groups. The first of these has order 60 and is the group Galois discovered in his work on the quintic. Then there are finite analogues of the four classical families of simple Lie groups, obtained by replacing the complex numbers by a finite field. Some of these were known to Galois and Jordan. In addition there are five curiosities, found by Emile-Léonard Mathieu in 1861 and 1873, and associated with some exceptional 'finite geometries'. This was the state of play at the turn of the century.

For a long time very little progress was made, although Leonard Dickson managed to obtain analogues of the exceptional Lie groups of dimensions 14 and 78 over finite fields. Obviously the other three exceptional Lie groups were crying out for similar treatment; but nobody could find the right way to make this idea work. Then in 1957 Claude Chevalley found a unified approach which produced, for any simple Lie group and any finite field, a finite simple group, said to be of *Lie type*. A variation, producing 'twisted' groups, was found by Robert Steinberg, Michio Suzuki, and Rimhak Ree. Suddenly order began to appear amid the chaos. There were the cyclic groups, the alternating groups, the groups of Lie type—and five *sporadic* groups, those found by Mathieu. This is very reminiscent of the list of simple Lie algebras: lots of nice families and a few black sheep. Would the pattern (or lack of it) repeat?

An important step towards the final solution was taken by Walter Feit and John Thompson in 1963, who verified a conjecture made in 1906 by William Burnside, to the effect that every finite simple group (other than cyclic) has even order. Their proof is 250 pages long and occupies an entire journal issue, and it set a pattern for work on finite simple groups (both in style and size). It opened up a completely new avenue of attack, because every group of even order has an element g such that $g^2 = 1$, and Richard Brauer had pioneered a method to exploit these special elements. Zvonimir Janko used this technique in 1965 to find a new sporadic simple group. Other sporadic groups made their entrance, associated with the names of Donald Higman, Charles Sims, John MacLaughlin, Michio Suzuki, John Conway, Dieter Held, Bernd Fischer, R. Lyons, Michael O'Nan, and Arunas Rudvalis. A few went through a brief period when it was believed that they existed, on circumstantial evidence, but nobody had actually constructed them. Lengthy computer calculations were often required to pin the existence problem down. The trickiest case was the biggest sporadic group, the *Monster*, whose order, expressed in prime factors, is

$$2^{46} \cdot 3^{20} \cdot 5^9 \cdot 7^6 \cdot 11^2 \cdot 13^3 \cdot 17 \cdot 19 \cdot 23 \cdot 29 \cdot 31 \cdot 41 \cdot 47 \cdot 59 \cdot 71.$$

There was some doubt whether even a large computer would be able to handle such a brute. In 1982, to everyone's astonishment, Robert Griess slew the monster with his bare hands—a feat that would do credit to St George. He constructed it as a group of rotations in 196883-dimensional space (the lowest dimension possible!).

With the successful completion of a programme of research laid down in the 1970s by Daniel Gorenstein, it is now known that there are exactly twenty-six sporadic simple groups. Together with the known families, they exhaust the list of finite simple groups. The complete proof totals at least 10,000 printed pages, and required contributions from hundreds of mathematicians. It is a remarkable triumph.

Having settled this long-standing and absolutely central problem, the group theorists are moving on. One question that has not yet been given a full answer is: 'Which groups can occur as Galois groups?' The best guess is 'all of them' but nobody has been able to prove it. Shafarevich managed to prove that all soluble groups are Galois groups, so attention has switched to the simple groups. Recently Thompson proved that the Monster occurs as a Galois group!

Symmetry of atoms

One of the applications of group theory is to crystal symmetry. The atoms in a crystal are arranged in a regular repetitive structure, or *lattice*, which has a great deal of symmetry. The symmetries include *point symmetries*, which measure how symmetrically the atoms are arranged around a given position in the lattice and *translational* symmetries, whereby the structure of the whole lattice repeats at fixed distances in particular directions. An infinite array of square tiles is a two-dimensional example. In fact, there are known to be exactly seventeen types of crystallographic symmetry group in the plane—or, more vividly, seventeen basic types of wallpaper pattern. In three dimensions there are exactly thirty-two point groups, which combine with translations to yield 230 distinct types of crystal symmetry. This classification is fundamental to solid state physics. In four dimensions there are 4,783 types of crystal; results for higher dimensions are mercifully incomplete.

Beyond groups

The classification of crystals shows that no crystal lattice can have 5-fold symmetry (like a pentagon): only 2-, 3-, 4-, and 6-fold symmetries can occur. But recently physicists have found a new state of solid matter, called a *quasicrystal*, in which 5-fold symmetry does occur. The results suggest that group theory is not the ultimate description of symmetry in nature. A quasicrystal does not have the traditional lattice structure of a crystal. That is, if you move it sideways, it looks different in fine detail. However, there is still a kind of long-range order to the structure: there are a small number of possible directions in which the bonds between atoms can point, and these are the same anywhere in the solid. Such patterns were originally discovered by Roger Penrose as a mathematical recreation. Penrose found two shapes of tile, kites and darts, which could cover the entire plane, but without any periodicity. Alan MacKay found three-dimensional versions of these tiles. The ideas were put on a firm mathematical basis by D. Levine and

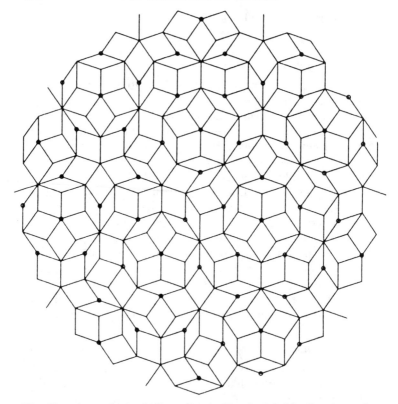

Fig. 13.   A quasicrystal tiling. (István Hargittai (ed.), *Quasicrystals, Networks, and Molecules of Fivefold Symmetry*, VCH Publishers.)

P. J. Steinhardt in 1984; and the predicted quasicrystal was observed in nature by Daniel Schechtman and others in an alloy of aluminium and manganese.

This discovery means that the traditional tool of crystallography, group theory, is not sufficient to deal with all 'symmetric' phenomena. Group theory has achieved dramatic successes, but it is not the last word. Mathematics must move with the times, it cannot rest forever on its laurels. The concept of symmetry may need to be broadened. Other 'forbidden' symmetries may arise in nature. And there are new challenges for mathematicians and physicists. 'Whatever the outcome,' says John Maddox, 'crystallographers will have to mug up some unfamiliar mathematics.' So will mathematicians.

10

Much ado about knotting

The chief contortionist of Barnum and Bailey's circus slapped me on the back and hailed me so boisterously that the embarrassed young lady who accompanied me made the error of attempting to continue her stroll as if nothing unusual were happening. But by the time she finally entered the big top, her circle of acquaintances had increased by three and she proudly held a handful of photographs, inscribed and autographed by the *Contortionist*, the *Tattooed Man*, and the *Expansionist*. 'Chippet', the contortionist, was none other than ex-'Bender' of my own defunct-but-never-absorbed-or-amalgamated circus, whom Barnum had succeeded in teaching to tie himself in a perfect FIGURE-OF-EIGHT KNOT. To this day I feel that P.T. had crowded me a bit.

Clifford W. Ashley

Until the end of the eighteenth century it was a common belief in Europe that magic knots could bewitch a bridegroom and prevent consummation of a marriage. In 1705 two people in Scotland were executed for using nine magic knots to interfere with the wedded bliss of a Mr Spalding of Ashintilly. Knots are symbolic of mystery. Although Shakespeare did not commit our chapter title, he did say, in *Twelfth Night*:

> O time! Thou must untangle this, not I;
> It is too hard a knot for me t'unite.

Mathematicians have found the problems posed by knots pretty hard t'untie too, and their higher-dimensional analogues pose even knottier problems. Only with the aid of time have they begun to untangle the strands of a general theory.

The mathematician is primarily interested in *topological* properties of knots—features that remain unchanged under continuous deformations. Knottedness itself is the most fundamental of these. Topology has been described as 'rubber sheet geometry', and this conjures up the right images even if it lacks technical accuracy. Knots are the most immediate topological features of curves in space. Beyond curves come surfaces; beyond surfaces come multidimensional generalizations called *manifolds*. Topologists study these, and a wild variety of other spaces. Topology is the study of the deeper influences of the notion of continuity, and considerations of continuity can turn up almost anywhere. As a result, topology has become one of the cornerstones of mathematics; and it is currently beginning to play a prominent role in applied science, especially mathematical physics (see chapters 15, 16, 17, 19). To those with a long-term historical outlook this is not so surprising, because many of topology's roots are in applications; but during its main developmental period, about 1920–70, the subject became very abstract, and many people lost sight of its potetial applicability.

Bridges and bands

Topology itself is generally agreed to have begun in 1735 with Euler's solution to the problem of the Königsberg bridges. The river Pregelarme has two islands. These are linked by a bridge. One island has one bridge crossing from it to each bank; the other has two bridges to each bank. Can the citizens of Königsberg cross all seven bridges once only in a continuous walk? Euler showed that they can't, and he solved the most general problem of the same type. His method is based on the observation that it is the way the bridges connect, not their precise positions or sizes, that matters.

There are also a few earlier discoveries in the 'prehistory' of topology. In 1639 René Descartes knew that if a polyhedron has V vertices, E edges, and F faces, then $V - E + F = 2$. Euler published a proof in 1751, and Cauchy gave another in 1811. Gauss said several times how important it was to study the basic geometric properties of figures, but he added little beyond a few observations on knots and links. A student of Gauss, Augustus Möbius, was the first person to define a

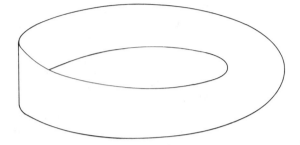

Fig. 14.   The Möbius band has a single side. If it is cut down the middle, it stays in one piece.

topological transformation as a one-to-one correspondence between figures, such that nearby points in one correspond to nearby points in the other. He studied surfaces and their relation to polyhedra, which later became a central theme in the subject. In 1858 he and Johann Listing discovered the possibility of one-sided (or *non-orientable*) surfaces, the most famous being the Möbius band. Take a long strip of paper, give it a half twist, and join the ends. If you try to paint what look like its two sides with different colours, the twist causes these to merge somewhere. If you slice it down the middle it stays in one piece.

The importance of holes

All this was amusing but of no clear significance. Topology became important for the rest of mathematics when Riemann introduced into complex analysis what are now called *Riemann surfaces*. Such a surface is equivalent to a sphere to which several handles have been added (just like handles on a cup), or to put it another way, a torus with several holes. The number *g* of handles or holes is called the *genus*.

Klein discovered that the projective plane (a sphere with antipodal points identified, as in the standard model of elliptic geometry) is non-orientable, and invented another surface, the *Klein bottle*, with this property. Neither surface can be placed in three-dimensional space without crossing itself, but each can be defined abstractly by identifying, or gluing, the edges of polygons. As a simple example, imagine a square. If the two

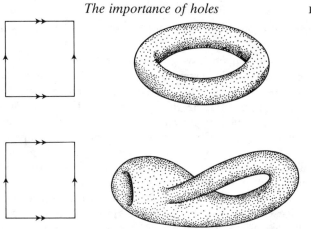

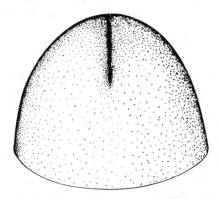

Fig. 15. The torus and Klein bottle can be constructed by abstractly identifying the edges of a square.

Fig. 16. The cross-cap is a surface, topologically equivalent to a Möbius band with a circular edge.

vertical edges are glued together it becomes a cylinder. If the top and bottom edges are now glued, the cylinder closes up into a torus. All properties of the torus can be studied using the *recipe* for the gluing, without actually going through the process of bending the square: it is sufficient to *pretend* that points due to be glued together are really the same. Now take a square; identify its two vertical edges; identify its two horizontal edges but in opposite directions. You've got a Klein bottle.

Conversely, given any surface whatsoever, you can slit it along some suitable system of curves and open the result up, like a peeled orange, to get a polygon. To reassemble the surface, you just imagine gluing the edges of the polygon back together. By studying how this ungluing and regluing procedure works, it is possible to classify completely the topological types of surfaces. The orientable ones are Riemann's g-handled spheres. The non-orientable ones are spheres in which g holes (with g at least 1) have been cut, and a Möbius band glued into each. This is possible in an abstract sense because the edge of a Möbius band is a single closed loop, that is, a topological circle. If you try to perform the construction in 3-dimensional space, the bands must cross themselves, forming hat-like objects called cross-caps. The surfaces just described are called the standard orientable and non-orientable surfaces of genus g. Any closed surface without edges is topologically equivalent to exactly one of them.

There is an important connection with Euler's Theorem. Although I didn't say so, the theorem that $V - E + F = 2$ is only true for polyhedra which are topological spheres—that is, which don't have any holes. If you make a polyhedral torus, you'll find that $V - E + F = 0$. For example a three-sided picture-frame built from wood with a triangular cross-section has $V = 9$, $E = 18$, $F = 9$. Now you can take a surface that interests you, draw a polyhedron on it, and calculate $V - E + F$. The fascinating thing is that the resulting value does not depend on how you draw the polyhedron: it depends only on the topological type of the original surface. It is a *topological invariant*, the *Euler number* χ. The genus is also a topological invariant: two topologically equivalent surfaces have the same genus. So is orientability/non-orientability. In fact the genus and the Euler number are related. For an orientable surface we have $\chi = 2 - 2g$, and for a non-orientable one $\chi = 2 - g$.

Knots

Knots are trickier than surfaces, even though they are only 1-dimensional. The reason is that when studying a knot we are not concerned just with the topological type of the curve (which is always a circle), but also *how it is embedded* in 3-space. Let me enlarge on this remark, because it's exactly here that the 'rubber-sheet' notion is misleading. A topological transformation is more general than a continuous deformation in some surrounding space. All that is required is to set up a one-to-one correspondence that keeps nearby points nearby. No surrounding space enters into this. To transform a knotted circle into an unknotted one is easy: just make corresponding points correspond! Another way to say it is that we first *cut* the knotted circle, then unknot it, then rejoin the cut exactly as it was. Although points near to each other across the cut become far apart during this process, they end up close together again; and that's all we need.

Two knots are embedded in 3-space in topologically equivalent ways if we can find a topological transformation of the *whole* of 3-space that sends one knot to the other. This is a far stronger requirement than merely making the knots correspond topologically. Let me abbreviate this to 'the knots are equivalent' to save breath. The main problems of knot theory are:

(1) Decide whether a knot really is knotted.
(2) Given two knots, decide whether they are equivalent.
(3) Classify all possible knots.

Until the 1920s it wasn't even a theorem that knots existed at all! It was proved by Hans Reidemeister that the ordinary overhand knot, or *trefoil*, is not equivalent to an unknotted loop. He did this by breaking up any deformation of a knot into a series of standard moves, and finding a property of the trefoil that is preserved by each such move, but which fails to hold for an ordinary circle. This was the first *knot invariant*.

A famous knot-theoretical Alexander, nicknamed 'The Great' was responsible in 333 BC for untying the immensely complicated *Gordian Knot*. He did this by a partial application of the 'cut and glue back' technique, only being a young man in a hurry he didn't bother about the gluing back, preferring to conquer the world rather than make his transformation topological. Another knot-theoretical Alexander, J. W., invented a beautiful knot invariant called the *Alexander Polynomial*. An unknotted loop has Alexander polynomial 1, the trefoil $x^2 - x + 1$, and these are different; *ergo* the knots are inequivalent. The figure 8 has polynomial $x^2 - 3x + 1$, so it's different from both. Sadly the reef knot and granny both have polynomial $(x^2 - x + 1)^2$, so the Alexander polynomial isn't powerful enough to yield a proof that generations of boy scouts can't be wrong. More sensitive invariants have recently been found: see chapter 20.

Conjurers often tie two separate knots in a rope, and manage to make the second 'cancel out' the first. Topologists know that this must be a trick, and I can't resist showing you a proof. Suppose knots K and L are tied in the same string (neatly separated from each other): call this the *connected sum* $K \# L$. Suppose a knot K has an 'inverse' K^* that cancels it out; and let 0 denote an unknotted loop. Tie infinitely many copies of K and K^* alternately on a single string, shrinking rapidly in size to keep the whole business finite. This gives an infinitely complicated knot

$$W = K \# K^* \# K \# K^* \# K \# K^* \# \ldots$$

Bracketing this expression one way, we have

$$\begin{aligned} W &= K \# (K^* \# K) \# (K^* \# K) \# \ldots \\ &= K \# 0 \# 0 \# \ldots \\ &= K. \end{aligned}$$

On the other hand,

$$\begin{aligned} W &= (K \# K^*) \# (K \# K^*) \# \ldots \\ &= 0 \# 0 \# 0 \ldots \\ &= 0. \end{aligned}$$

So $K = 0$, that is, K is unknotted! Perhaps the biggest surprise is that this ridiculous argument can easily be made perfectly rigorous. Infinite knots like W are called *wild knots*. A mathematician called R. L. Wilder invented a particularly nasty class of wild knots: you can guess what these are called. The connected sum operation # produces a sort of knot arithmetic. Define a knot to be *prime* if it isn't the connected sum of two other knots (other than 0 and itself). Then every knot can be written *uniquely* as the connected sum of prime knots.

As well as knotted curves, it is possible to knot surfaces. As the Danish mathematician Piet Hein wrote in one of his 'grooks' (epigrammatic poems):

> There are doughnuts and doughnuts
> With knots and with no knots
> Any many a doughnut
> So nuts that we know not.

Manifolds

The generalization of topology from surfaces to higher dimensions was initiated by Riemann, although he was thinking more in terms of rigid geometric properties. He introduced the idea of a *manifold*.

I'd better start by explaining what mathematicians mean by n-dimensional space. The idea is that the plane can be thought of as the set of points with coordinates (x, y) where x and y are numbers; and 3-space in terms of points (x, y, z). So 4-space is the set of points (x, y, z, u), 5-space (x, y, z, u, v) and so on. That's it: nothing could be simpler! And it only took 3,000 years or so to see how obvious it all is.

A surface, however curved, however complicated, can be thought of as lots of little round patches glued together; and topologically each patch is just like a patch in the ordinary Euclidean plane. It is not this *local* patch-like structure that produces things like the hole in a torus: it is the *global* way all the patches are glued together. Once you see this, the step to n dimensions is easy: just assemble everything from little patches cut out of n-dimensional space instead of the plant. And that, aside from a few technical points which nobody sorted out until the 1930s, is exactly what is done. The resulting space is an *n-dimensional manifold*. Usually this is abbreviated to 'n-manifold'. There are good reasons to want to know about manifolds. For example the motion of three bodies under gravity is really a question about an 18-dimensional manifold, with three position coordinates and three velocity coordinates per body.

What topologists would like to do to manifolds is what they're already done for surfaces and knots. Namely:

(1) Decide when two manifolds are or are not topologically equivalent.

(2) Classify all possible manifolds.
(3) Find all the different ways to embed one manifold in another (like a knotted circle in 3-space).
(4) Decide when two such embeddings are, or are not, the same.

The first two problems are bad enough; the second two worse. But a great deal of progress has been made. I'll concentrate on (1) and (2); but I'd like to mention one result which shows that (3) must be pretty messy. The question is: when can two 50-dimensional spheres be *linked* in n-dimensional space? By 'link' I mean that they can't be separated by a topological transformation of the surrounding space. The answer is that they cannot link if n is 102 or more; they can link if $n = 101$, 100, 99, 98; but—here comes the surprise—they cannot link if $n = 97$ or 96! Then they can link again if $n = 95$, and so it goes down to $n = 52$. Weird. Topology is full of counter-intuitive things like that.

The Poincaré Conjecture

Until very recently progress on problems (1) and (2) above was stymied by a difficulty that Poincaré ran into in 1904, saying rather offhandedly 'this question would lead us too far astray' before proceeding rapidly to other things. He was right: after eighty years of being led astray nobody knows the answer to his exact question, although higher-dimensional versions have been settled completely.

The problem lies in an area called *homotopy theory*. This is a part of *algebraic topology*. The idea in algebraic topology is to reduce topological questions to abstract algebra by associating with topological spaces various algebraic invariants. Poincaré was one of the fathers of this theory, and he invented a thing called the *fundamental group*. The idea is a cunning blend of geometry and algebra. Take a manifold M and select a point * in it. Consider loops that start and end at *. You can 'multiply' two loops by doing first one, then the other: the result is again a loop starting and ending at *. However, you can't 'divide' loops. This can be achieved, though, if you pretend that any loop is the same as a continuous deformation of itself. The reciprocal, or inverse, of a loop is then the same loop traversed in the opposite direction. The set of all loops, with this 'multiplication', is Poincaré's fundamental group. If M is a circle then the thing that distinguishes loops is how many times they wind round; and the fundamental group consists of the integers. The 'multiplication' is actually addition: if the first loop winds 5 times and the second 7, then the two together wind 12 times, not 35.

After Poincaré's lead, other mathematicians noticed that if you use k-dimensional spheres instead of 1-dimensional loops you get *higher homotopy groups*, one for each k. Each higher homotopy group is a topological invariant, in the sense that two topologically equivalent manifolds must have the *same* homotopy groups. Now Poincaré wondered whether these invariants are good enough to tell any two different manifolds apart. In particular he asked: if M has the same homotopy groups as the 3-dimensional sphere S^3, then is M topologically equivalent to S^3? His guess, 'yes', remains unproved to this day. The problem has also been generalized by later workers. The n-dimensional Poincaré Conjecture is that an n-dimensional manifold with the same homotopy groups as the n-dimensional sphere S^n is topologically equivalent to S^n. It is easy to check that this is true for $n = 2$, by using the classification of surfaces. But in higher dimensions no classification of manifolds is known, which puts a different complexion on finding a proof.

Five . . .

In 1961 Stephen Smale developed a method for breaking a manifold up into nice pieces, called *handles*. By cleverly moving handles around he proved an important result, called the h-cobordism theorem, in dimensions $n \geq 5$. Among its consequences is the Poincaré Conjecture for dimensions 5 or higher. In fact Smale's proof first worked only for $n \geq 7$. John Stallings dealt with the case $n = 6$ and Christopher Zeeman with $n = 5$. Then Smale extended his methods to cope with those cases too. However, an essential ingredient, the 'Whitney trick', fails completely in dimensions 3 or 4, and there seemed to be no way to extend Smale's proof to these cases. It became the conventional wisdom that topology is somehow different in dimensions 3 and 4, mostly because nobody could prove anything really useful in those dimensions.

Four . . .

In 1982 Michael Freedman found a way to get round the failure of the Whitney trick, and showed that the conventional wisdom was wrong. In four dimensions, manifolds behave pretty much the way they do in five or more. In particular, the Poincaré Conjecture is true for 4-dimensional manifolds.

Three . . .

This meant that the only unsettled case of the Poincaré Conjecture was the one originally posed by Poincaré, in three dimensions. This tantalizingly narrow gap happens because 2-dimensional space is too small to have room for any serious complexity, and 4- or higher dimensional space is so big that the complexities can be rearranged nicely. In three dimensions there is a creative tension: big enough to be complicated; too cramped to be easily simplified. What was needed was a line of attack that exploited the special properties of 3-dimensional manifolds. But what are those special properties? Topologists joined the hunt, moving into 3-manifolds in droves.

Rigidity in floppiness

A topological manifold is a very floppy thing, free to be kneaded and distorted virtually at will, like a lump of pizza dough. Traditional geometry, Euclidean and non-Euclidean, is much more rigid, like a cooked pizza crust. There is a metric, a distance structure, and distances cannot be altered. Stretching is not allowed. So it was a great surprise when in the mid-1970s William Thurston announced a programme to reduce the study of 3-manifolds to geometry!

The idea goes back to the nineteenth-century work on surfaces. One way to sum up the classical results on non-Euclidean geometry is to say that, up to a choice of scale, there are three basic geometries, each realizable as the geometry of a surface of constant curvature. The curvature distinguishes the three cases:

(1) the Euclidean plane, of curvature 0;
(2) the hyperbolic plane, of curvature -1;
(3) the elliptic plane or 2-sphere, of curvature $+1$.

Now it turns out that every topological type of surface is associated with, and can be constructed from, exactly one of these geometries. Specifically, each surface can be given a metric of constant curvature 0, −1, or +1. Which one depends on a topological invariant: the Euler number.

There are only two surfaces with positive Euler number: the sphere and the projective plane. The sphere already has curvature +1; and since the projective plane is obtained from the sphere by identifying antipodal points, this metric can be transferred to the projective plane. (In fact this is just the usual model of elliptic geometry.) There are also two surfaces with Euler number zero, namely the torus and the Klein bottle. We saw above how to get these by gluing the edges of a square. Now a square lives in Euclidean space, so inherits its zero-curvature metric; therefore the torus and Klein bottle inherit it too. Finally there is an infinite family of surfaces with negative Euler number. These can all be constructed by gluing the edges of a polygon in the hyperbolic plane! So they inherit its metric of curvature −1.

Given this beautiful relation between geometry and topology, it ought to be natural to ask whether it generalizes to three dimensions. However, there are several complications. As Thurston says: 'Three-manifolds are greatly more complicated than surfaces, and I think it is fair to say that until recently there was little reason to suspect an analogous theory for manifolds of dimension 3 (or more)—except perhaps for the fact that so many 3-manifolds are beautiful.' But he then adds, 'The situation has changed.'

Eight geometries

One of the complications is that instead of the three geometries possible in two dimensions, there are *eight* possible geometries in three dimensions, as Thurston showed in 1983. Naturally, they include Euclidean 3-space, hyperbolic 3-space, and elliptic 3-space (or the 3-sphere); but there are five others. The proof that only these eight occur requires a lengthy analysis, based on group theory. This should be no surprise, given Klein's Erlangen programme: geometry is group theory! All sorts of familiar theorems make guest appearances in the proof, for example the classification of the seventeen wall-paper groups.

Just as surfaces can be given metrics modelled on the three types of two-dimensional geometry, so *some* 3-manifolds can be given metrics modelled on one of the eight geometries above. Each 3-manifold corresponds to at most one type of geometry. However, some 3-manifolds have no such geometric structure at all.

The way to get round this difficulty, according to Thurston's programme, is to break the manifold up into pieces, and put a geometric structure on each bit. The conjecture is that this can be done in only one way. First split it into prime manifolds (much as knots can be decomposed into prime knots via the connected sum #), then cut along specially chosen tori. There is a lot of evidence in favour of this proposal, and none against. In fact, for seven of the eight geometries, the problem has been solved completely.

The hyperbolic phoenix

That leaves just one final case, which happens to be 3-dimensional hyperbolic geometry; but that seems to be a tougher nut to crack. It must be, because the 3-dimensional Poincaré Conjecture will appear as a minor detail if Thurston's programme can be completed. The reason people are excited about the programme is that it places the problem within a much richer framework, adding a lot of leverage. The older theory of knots and links is important for the study of 3-manifolds, because the complement of a knot (that is, the set of points *not* on the knot) gives rise to a 3-manifold. The hyperbolic geometry of knot complements is a rich and beautiful area.

One by-product of Thurston's programme is a proof of the Smith Conjecture, stated by Paul Smith in the 1940s. This is about transformations of a 3-sphere into itself, which are *periodic* in the sense that if you perform them a suitable number of times, everything ends up where it started. Some points never move at all, and the conjecture is that the set of all these *fixed points* (which is known to be a curve) isn't a knot. The proof, completed in 1978, assembles the results of a number of people, among them Thurston, Hyman Bass, William H. Meeks III, and Shing-Tung Yau.

One effect of all this is a sudden renewal of interest in hyperbolic geometry, which had been thought of as pretty much mined out. In fact little work was ever done on the 3-dimensional case, but only now does it appear sufficiently interesting and relevant to the mainstream to be worth doing. As the needs of research mathematics change, supposedly 'dead' subjects can rise like the phoenix from their own ashes.

Three kinds of glue

My remark above, that 4-dimensional manifolds are much like their higher-dimensional cousins, isn't completely true. It depends on what brand of manifold you buy. Although I have concealed this fact so far, there are three kinds of manifolds: piecewise-linear, topological, and smooth. Up till now the difference hasn't really been important, but I want to talk about one of the most sensational discoveries ever made in topology, where the difference is the essence of the matter.

As I said above, you can think of a surface as being cut into a number of pieces, each of which can be 'flattened out' by a topological transformation to look like a piece of the plane. The original surface can be built up by overlapping all of these pieces. Indeed this is how mathematicians define a surface in the abstract: a lot of pieces of the plane, 'glued together' on their overlaps. A similar approach is used in n dimensions, but now the pieces are taken from Euclidean n-space. The way the gluing takes place determines what type of surface you have, and there are—so to speak—three important types of glue. Real glue joins each point on one object to a point on the other. Mathematical glue just specifies, for each point on one, the corresponding point on the other (to which the real glue would attach it). That is, it is determined by a one-to-one correspondence between the overlapping parts. The type of glue is determined by what properties this correspondence preserves.

If you use topological glue, only notions of continuity are preserved. Points nearby on one piece must be glued to points that are nearby on the other piece; but you can bend the pieces however you like. If you use smooth glue, then smooth curves on one piece must be glued to smooth curves on the other. You can't, for example, put a sharp corner in one piece and glue that to a straight line in the other. It is this notion of 'manifold' that Riemann had in mind. Finally, if you follow in the footsteps of Poincaré and use piecewise-linear glue, then straight lines on one piece must glue to straight lines on the other. An irregular lump of dough is a sphere with a topological structure; and egg is a sphere with a smooth structure; and a wedge of cheese is a sphere with a piecewise-linear structure. The distinction is technical (essentially it lets you use different implements from the mathematical toolkit, according to your tastes), but the technicalities here are important, and in higher dimensions make a big difference. A theorem may be true for piecewise-linear manifolds, but not for smooth manifolds, and so on. It's important to understand how these relations go, because it may be easier to prove theorems for (say) the piecewise-linear case, and then pass to the topological or smooth cases by some standard trick. (For example in the 1930s Henry Whitehead discovered how to give every smooth manifold a piecewise-linear structure—that is, how to bend an egg until it looks like a wedge of cheese. And in the 1950s it was shown that every topological 3-manifold has a *unique* smooth structure—that is, every lump of dough can be smoothed out into an appropriate egg shape. The 'egg' can have handles and holes and knots and so on, of course—it's the smoothness that's important.) In mathematical warfare, it's good strategy to choose the battleground that best suits your weaponry. A large part of the 1960s research effort in topology centred around these relationships, culminating in the results of Robion Kirby and Larry Siebenmann which sorted everything out in dimensions 5 or higher—in a nice way, but with the odd surprise.

Fake 4-space

One absolutely basic question is the uniqueness of the smooth structure on a given smooth manifold. In other words, is there only one consistent way to define 'smoothness' on a given space? Originally, everyone pretty much assumed the answer would be 'yes'; but Milnor's exotic 7-sphere (mentioned in chapter 1) showed that it is 'no'. Indeed there are exactly 28 distinct smooth structures on a 7-sphere. However, it was still hoped that for good old *Euclidean* space of dimension n there should be a unique smooth structure, namely the standard one used in calculus. In fact this was known to be true for all values of n except 4. But at much the same time that Freedman was showing that 4-dimensional space is in most respects pretty much like five or higher dimensions, Simon Donaldson, in 1983, made the dramatic discovery of a non-standard smooth structure on 4-space.

His proof is difficult and indirect, making essential use of some new ideas cooked up between them by topologists and mathematical physicists called *Yang–Mills gauge fields* (see chapter 19). These were invented to study the quantum theory of elementary particles, and were not intended to shed light on abstruse questions in pure topology! On the other hand, the topological ideas involved were not invented for the analysis of fundamental particles either. It is quite miraculous that two such disparate viewpoints should become united in such a tight and beautiful way. Ironically, many of Freedman's ideas are essential to the proof; and the starting-point is a curious group-theoretical coincidence. The Lie group of all rotations of n-dimensional space is simple, except when $n = 4$. In that case it breaks up into two copies of the rotation group in 3-space. That such a coincidence should have such stunning significance is yet another triumph for the Erlangen programme.

At any rate, we now know that 4-dimensional space is quite unusual, compared to any other dimension. In fact, not only is the smooth structure non-unique: it is *extremely* non-unique. Milnor's 7-sphere has only a finite number of smooth structures. But there is an *uncountable infinity* of distinct smooth structures on 4-space. However, things may not be quite as bad as they currently seem. Among these is one 'universal' structure which in a sense contains all the others. Perhaps the issues will eventually be resolved so that everything appears natural and inevitable; but right now there's still plenty of mystery. In particular, it would be nice to have a good answer to one very basic question: what do these exotic smooth structures on 4-space *look* like?

11

The purple wallflower

If in any process there are a number of contradictions, one of them must be the principal
contradiction playing the leading and decisive
role, while the rest occupy a secondary and
subordinate position. Therefore, in studying any
complex process in which there are two or more
contradictions, we must devote every effort to
finding its principal contradiction. Once this
principal contradiction is grasped, all problems
can be readily solved.

Mao Tse-tung

In 1852 Francis Guthrie, a graduate student at University
College London, wrote to his younger brother Frederick with
a simple little conundrum that he had stumbled across while
attempting to colour a map of the counties of England. 'Can
every map drawn on the plane be coloured with four (or
fewer) colours so that no two regions having a common border
have the same colour?'

'Map' here is used in the geographical sense: a division
of the plane into finitely many regions bounded by curves.
'Common border' refers to a definite line, and not just an
isolated point, along which the regions meet. Frederick
Guthrie couldn't solve it, so he asked his professor, the distinguished mathematician Augustus De Morgan. On 23 October
1852 De Morgan confessed in a letter to the even more distinguished William Rowan Hamilton that he couldn't make
any headway. This is the earliest documented reference to the
Four Colour Problem, which became notorious as the most
easily *stated*, but most difficult to answer, open question in
mathematics. It remained unsolved until 1976. Appropriately
enough it has a colourful history, and presents subtle traps for
the unwary. Hermann Minkowski, a major figure of the nine-

teenth century, once told his students that the only reason it
hadn't been solved was that only third-rate mathematicians
had attacked it. 'I believe I can prove it', he announced. Some
time later he told them sheepishly, 'Heaven is angered by my
arrogance; my proof is also defective.' As Thomas Saaty and
Paul Kainen remark: 'One of the many surprising aspects of
the Four Colour Problem is that a number of the most important contributions to the subject were originally made in the
belief that they were solutions.'

The eventual positive answer is unusual in at least a technological sense. The proof depends on showing that some
2,000 specific maps behave in a particular way. Checking
all these cases is immensely tedious, and requires several
thousand hours of time on a fast computer. If a full proof were
to be written out, it would be so enormous that nobody could
live long enough to read it—let alone verify it. Is such a
monstrosity *really* a proof? Opinions are divided. Does a
simpler proof exist? Nobody knows, although it *has* been
shown that no substantially simpler proof can run along similar
lines.

Despite its notoriety, the Four Colour Problem is not really
in the mainstream of mathematics: its solution doesn't seem to
open up any areas of major significance. It is more of a
tidying-up exercise. But its solution introduced some novel
ideas, requires little specialist technique to follow, and poses
some intriguing philosophical questions which shed a lot of
light on the way mathematicians feel about their subject. No
discussion of the problems of mathematics would be complete
without giving it a mention.

Comedy of errors

De Morgan did manage to show that there is no way to
arrange five countries so that any two are adjacent: this *suggests* that four is the maximum number of colours needed,
but it isn't a proof because there might be other 'non-local'
obstacles to 4-colourability, such as complicated chains of
regions linking up in funny ways. In 1878 came the first reference to the problem in print, a letter by Cayley in the *Proceedings of the London Mathematical Society*, asking whether
the problem had been solved yet. In 1879 Arthur Kempe, a

barrister and a member of the Society, published a proof, and the matter seemed settled. It was a minor curiosity, nothing more.

The proof involved an ingenious use of what are now called *Kempe chains*. Suppose for definiteness that the desired colours are red, blue, yellow, and green. Pick a pair, say red and yellow. Suppose that a portion of the map has successfully been coloured. Think of it as a number of red-yellow 'islands' in a blue-green 'sea'. (There may be lakes on islands and islands within lakes.) Observe that the colours red and yellow may be interchanged throughout any individual island without creating any forbidden adjacencies, because internal borders remain red-yellow, while external borders are either unchanged, or are on the 'coast', with sea (blue-green) one side and land (red-yellow) on the other. You can play this game with any pair of colours. Each island is a Kempe chain; and the possibility arises of colouring a map until you get stuck, and then changing colours in carefully chosen Kempe chains to get unstuck again and continue.

Minimal criminals

More formally, we resort to a standard mathematical ploy, 'proof by contradiction'. If the Four Colour Theorem is *false*, then there must exist maps that require a fifth colour—purple, say. If such 'bad' maps exist, they can be incorporated into bigger maps in all sorts of ways, all of which will need purple passages too. Since there's no fun in making bad maps bigger, we go the opposite way, and look at the smallest bad maps. These *minimal criminals* need five colours; but any *smaller* map needs no purple regions. The aim is to use this information to restrict the structure of a minimal criminal, to exploit these restrictions to pin it down further, and so on—until eventually it is shown that no such creature exists. No petty crime implies no big crime either . . . so the theorem is true after all!

Kempe's idea was to take a minimal criminal, and shrink a suitable region to a point. The resulting map has fewer regions, so can be 4-coloured. Restoring and colouring the shrunken region may not be possible, because it may abut regions that between them already use up all four colours. If it's a triangle (a region meeting only three others) that's not a problem. If it's a square, then a Kempe chain swap can change one neighbouring colour, which does the trick. If it's a pentagon, says Kempe, a similar argument works. And he could prove that *every* map must contain either a triangle, a square, or a pentagon.

But the Four Colour Problem wasn't going to succumb that easily. In 1890 Percy Heawood found a mistake in Kempe's treatment of pentagonal regions. A year later P. J. Tait published a different proof, but J. Petersen quickly found a mistake in that too. Heawood did notice that Kempe's method can be patched up to give a proof that *five* colours always suffice. (One extra colour makes the pentagon easier to restore.) He also studied analogues of the problem on surfaces other than the plane, giving a bound for the number of colours required, in terms of the number of holes in the surface. The *Heawood Conjecture*, that this number is the best possible bound, was proved in 1968 by G. Ringel and E. Youngs (using some beautiful and ingenious combinatorial ideas). There were two exceptions. One was the sphere (or plane) where the Heawood bound of 4 remained conjectural, The other was the Klein bottle, where the Heawood bound is 7 but the correct answer is known to be 6. So by 1968 the correct number was known for all surfaces *except* the plane! This illustrates a possibly undesirable tendency in mathematics: to do a lot of work on generalizations without solving the original problem. Of course, the generalizations might lead to an idea that solves everything: they often do. This time they didn't.

Shrink and restore

Back to the original Four Colour Problem. In 1922 Philip Franklin proved that every map with 26 or fewer regions is 4-colourable. His method laid the foundations for the eventual successful assault, with the idea of a *reducible configuration*. A *configuration* is just a connected set of regions from the map, together with information on how many regions are adjacent to each around the outside. To see what reducibility means, let's take an example. Suppose a map contains a triangle. Shrink it to a point. Suppose the resulting map, which has one region fewer, can be 4-coloured. Then so can the original map.

The triangle abuts only three regions, leaving a fourth colour spare when it is restored to the map. In general, a configuration is reducible if the 4-colourability of any map that contains it can be proved provided a smaller map is 4-colourable. (The 'shrink-colour-restore-extend' strategy can be varied in all sorts of ways, but is the basic way to prove reducibility.) A similar argument proves squares are reducible.

A minimal criminal cannot contain a reducible configuration. For suppose it did. Reducibility implies that, if a certain smaller map is 4-colourable, then so is the original. But minimality means that the smaller map *is* 4-colourable, so the minimal criminal is too. And that contradicts its nefarious nature. What we must do, then, is show that all minimal criminals must contain some reducible configuration. That way, we rule them all out. The most direct way to do this is to find a set of reducible configurations which is *unavoidable*, in the sense that any map must contain a configuration in this set. Kempe effectively tried to do this. He proved, correctly, that the set {triangle, square, pentagon} is unavoidable. But he made an error when proving reducibility of the pentagon. Nevertheless, the basic strategy of a proof now emerges: find an unavoidable set of reducible configuations. The list will have to be longer than Kempe's, because replacements for the pentagon will be needed.

Discharging

In 1950 Heinrich Heesch became the first mathematician to state publicly that he believed the Four Colour Theorem could be proved by finding an unavoidable set of reducible configurations. He estimated that about 10,000 configurations would be needed, each of moderate size. Heesch and his student Karl Dürre developed a computer program to test for reducibility. Like its more sophisticated successors, it wasn't foolproof. If the test was successful, the configuration was reducible; but if it failed, reducibility remained open.

Heesch also devised a method for proving unavoidability, based on a loose electrical analogy. Suppose a quantity of electrical charge is applied to each region, and then allowed to move into neighbouring regions by following various rules. For example, we might insist that the charge on any pentagon is split into equal parts and transferred to any of its neighbours, except triangles, squares, and pentagons. By analysing the general features of charge distributions, it can be shown that certain specific configurations must occur—otherwise charge 'leaks away'. More complicated recipes lead to more complicated lists of unavoidable configurations.

Man and machine

In 1970 Wolfgang Haken found improvements to Heesch's discharging method and started thinking seriously about solving the Four Colour Problem. The main difficulty was the likely size of configurations in the unavoidable set. The *ring size* of a configuration is the number of regions that surround it. There was reason to believe that a ring size of 18 or more might be needed. As well as not being foolproof, reducibility tests grow exponentially with the ring size, increasing by a factor of about 4 for every extra region in the surrounding ring. While an 11-ring configuration isn't too bad, a 14-ring configuration will take 26 hours without doing any of the fancy extras likely to be needed. Even if that can be reduced to about half an hour on average, most 18-ring configurations will take 100 hours or more—and the calculations will use more memory storage than any computer possesses. With an estimated 10,000 regions in an unavoidable set, the whole computation could take a century. And if, at the end, just one configuration in the unavoidable set turns out not to be reducible, then the whole calculation is worthless! The computer chases its tail and succeeds only if it catches it. So between 1972 and 1974 Haken, together with Kenneth Appel, began an interactive dialogue with the computer, to try to improve the chances.

The first run of their computer program produced a lot of useful information. Some of it was positive—the likely ring size dropped to 16; the likely size of an unavoidable set was smaller too. Some was not: the quantity of computer output would be too big to handle, and there were flaws in the method. Thus armed, they modified the program to overcome the flaws and tried again. More subtle problems emerged and were duly corrected. After some six months of this dialogue, Appel and Haken became convinced that their method of proving unavoidability had a good chance of success. It took another year to work out a proof of this. In 1975 they modified the program again. They write:

At this point the program began to surprise us. At the beginning we would check its arguments by hand so we could always predict the course it would follow in any situation; but now it suddenly started to act like a chess-playing machine. It would work out compound strategies based on all the tricks it had been 'taught', and often these approaches were far more clever than those we would have tried. Thus it began to teach us things about how to proceed that we never expected. In a sense it had surpassed its creators in some aspects of the 'intellectual' as well as the mechanical parts of the task.

The home straight

By 1975 there seemed to be a real chance that the method would prove practicable, and the research programme moved from the exploratory phase to the final attack. The first step was to write an efficient program to prove reducibility, using assembly code on an IBM 360 at the University of Illinois. Much of the programming was done by John Koch, a computer scientist. The discharging procedure used to prove unavoidability had by now been modified so much that, rather than rewriting the program, it was decided to do that part of the work by hand. Jean Mayer, a professor of French Literature at Montpellier with a lifelong interest in the Four Colour Problem, suggested an important improvement in the procedure, and the likely ring size dropped to 14. In January 1976 construction of the unavoidable set began. Again it proceeded interactively, using the computer to check ideas and results. By June 1976 the work was complete. The computer duly reported that every configuration in Appel and Haken's unavoidable set was reducible. Red, blue, yellow, and green alone suffice to colour a planar map: the purple wallflower need never be invited to join the dance.

Is that *a proof?*

To what extent can an argument that relies on an enormous and humanly uncheckable computation really be considered a proof? Stephen Tymoczko, a philosopher, wrote: 'If we accept the four-colour theorem as a theorem, then we are committed to changing the sense of "theorem", or more to the point, to changing the sense of the underlying concept of "proof".' The curious thing is that few mathematicians agree. This is so even though the correctness of the computer calculations requires a definite act of faith: it is not something that, even in principle, anybody could check.

There exist mathematical proofs that are so long and complicated that, even after studying them for a decade, nobody could put his hand on his heart and declare them to be totally unflawed. For example the classification theorem for finite simple groups is at least 10,000 pages long, required the efforts of at least a hundred people, and can be followed only by a highly trained specialist. However, mathematicians are generally convinced that the proof of the Four Colour Theorem is correct. The reason is that the strategy makes sense, the details hang together, nobody has found a serious error, and the judgement of the people doing the work is as trustworthy as anybody else's. Humans, by their nature, make mistakes. But serious mistakes in mathematics usually show up sooner or later, because the subject is constantly cross-checking results in one area against results in another. About a decade ago two teams of topologists, one Japanese and the other American, got exactly opposite answers to an important question. It took a while to sort out which team was wrong, and the Americans got annoyed because nobody really seemed very bothered; but eventually the mistake was found.

What mathematicians really like is an *illuminating* proof. That's matter of taste rather than logic or philosophy. Davis and Hersh make a remark that must have flashed through many mathematicians' minds when the proof was announced:

My first reaction was, 'Wonderful! How did they do it?' I expected some brilliant new insight, a proof which had in its kernel an idea whose beauty would transform my day. But when I received the answer, 'They did it by breaking it down into thousands of cases, and then running them all on the computer, one after the other,' I felt disheartened. My reaction was, 'So it just goes to show, it wasn't a good problem after all.'

There is nothing in the Appel–Haken proof that is any less *convincing* than, say, the classification of simple groups; or indeed of the hosts of theorems published every day that are outside a particular mathematician's speciality—which he will never check himself. It is also much less likely that a computer will make an error than a human will. Appel and Haken's proof strategy makes good logical sense and is sufficiently standard that no fallacy is likely there; the unavoidable set was in any case obtained by hand; and there seems little reason to doubt the accuracy of the program used to check reducibility. Random 'spot tests' and, more recently, repetitions of the computations using different programs, have found nothing amiss. I know of no mathematician who currently doubts the correctness of the proof.

Davis and Hersh also remark:

To the philosopher, there is all the difference in the world between a proof that depends on the reliability of a machine and a proof that depends only on human reason. To the mathematician, the fallibility of reason is such a familiar fact of life that he welcomes the computer as a more reliable calculator than he himself can hope to be.

Perhaps we can leave the last word to Haken, who said in a newspaper interview:

Anyone, anywhere along the line, can fill in the details and check them. The fact that a computer can run through more details in a few hours than a human could ever hope to do in a lifetime does not change the basic concept of mathematical proof. What has changed is not the theory but the practice of mathematics.

12
Squarerooting the unsquarerootable

The Divine Spirit found a sublime outlet in that
wonder of analysis, that portent of the ideal
world, that amphibian between being and not-
being, which we call the imaginary root of
negative unity.

Gottfried Leibniz

The seventeenth century witnessed a new flowering of algebra,
with the solution of cubic and quartic equations as the centre-
piece. But still nobody was sure what a *number* was. Negative
numbers were still causing headaches in Europe, even though
the Hindu Brahmagupta had dealt satisfactorily with them in
AD 628. In the mid-1600s Antoine Arnauld argued that the
proportion $-1:1 = 1:-1$ must be nonsense: 'How can a
smaller be to a greater as a greater is to a smaller?' The
dangers of verbal reasoning in mathematics could hardly be
plainer; yet in 1712 we find Leibniz agreeing that Arnauld had
a point. However, Thomas Harriot (1560–1621) worked with
negative numbers, and Raphael Bombelli (1526–73) even gave
clear definitions. Simon Stevin and Albert Girard treated
negative numbers as being on a par with positive ones.

But even Brahmagupta couldn't swallow *square roots* of
negative numbers; so the sparks were bound to fly when, as
Morris Kline puts it, 'without having fully overcome their
difficulties with irrational and negative numbers the Europeans
added to their problems by blundering into what we now call
complex numbers.' Thus in 1545 we find Girolamo Cardano,
one of the great scoundrels of mathematics and a genius to
boot, solving $x(10 - x) = 40$ and emerging with the solutions
$5 + \sqrt{-15}$, $5 - \sqrt{-15}$. 'Putting aside the mental tortures
involved', says he, substitute the supposed answer in the
equation. Lo and behold, you find that

$$x(10 - x) = (5 + \sqrt{-15})(5 - \sqrt{-15}) = 5^2 - (\sqrt{-15})^2$$
$$= 25 - (-15) = 40$$

as required. 'So', says Cardano, 'progresses arithmetic subtlety, the end of which is as refined as it is useless.'

A phrase that should haunt mathematicians is: 'it doesn't make sense but it works.' Coincidences like Cardano's are akin to the visible tracks left in the jungle by some passing creature: a sign that some wondrous discovery lies in wait for a sufficiently intrepid adventurer. So *does* −1 have a square root? If so, what manner of beast is it? Why should practical minds concern themselves with such otherworldly intellectual frippery?

Ars Magna

Cardano's manipulations occurred in his *Artis Magnae*, a celebrated work summing up Renaissance algebra, some of it pinched from others and published with due acknowledgement but without their permission. In the same book he noticed that Tartaglia's formula for the cubic (chapter 1) can lead into murky waters. Applied to the equation $x^3 = 15x + 4$ it yields the answer

$$x = \sqrt[3]{2 + \sqrt{-121}} + \sqrt[3]{2 - \sqrt{-121}}.$$

Since the obvious answer is $x = 4$, something is clearly rotten in the state of Bologna. Bombelli, who appears to have had an unusually clear head, observed that (again putting aside any qualms)

$$(2 \pm \sqrt{-1})^3 = 2 \pm \sqrt{-121},$$

so Cardano's expression becomes

$$x = (2 + \sqrt{-1}) + (2 - \sqrt{-1}) = 4,$$

and the idiotic formula is actually telling us something sensible after all. Bombelli had the first inkling that what we now call *complex* numbers can be useful for getting real results.

Descartes in 1637 called expressions involving square roots of negative numbers 'imaginary', and took their occurrence as a sign that the problem was insoluble. Isaac Newton shared the same opinion. So much for Bombelli. Nevertheless, mathematicians couldn't keep their sticky fingers off the goodies. Having floundered in the quagmire of algebra they plunged headlong into the flaming pit of analysis. John Bernoulli, trying to integrate functions like $1/(x^2 + 3x + 1)$, stumbled over *logarithms* of complex numbers. By 1712 Bernoulli and Leibniz were engaged in a ding-dong battle over a far simple problem, logarithms of negative numbers. Bernoulli argued that since $d(-x)/-x = dx/x$ then integration shows that $\log(-x) = \log(x)$. Leibniz countered that the integration only works for positive x. Euler said they were both wrong, because integration requires an arbitrary constant, so the correct conclusion is that $\log(-x) = \log x + c$ for some constant c, obviously equal to $\log-1$, whatever *that* is. All was grist to Euler's mill, which in 1748 ground out the miraculous formula

$$e^{i\theta} = \cos\theta + i\sin\theta$$

where $i = \sqrt{-1}$. Actually it wasn't new: Roger Cotes had found much the same result in 1714, but this had been overlooked. Putting $\theta = \pi$ we get $e^{i\pi} = -1$, whence $\log(-1) = i\pi$. However, by cubing we also find that $e^{3i\pi} = -1$, so $\log(-1) = 3i\pi$. A lesser man would have given up in disgust, but Euler found a loophole: $\log x$ has many values. If L is one of them, then the others are $L \pm 2\pi i$, $L \pm 4\pi i$, $L \pm 6\pi i$, and so on. The jungle creature was no longer breaking just the odd twig, it was laying waste to great swathes of forest; more a herd of rhino than a squirrel. But could it be trapped?

Plane thinking

John Wallis, in his *Algebra* of 1673, interprets a complex number as a point in the plane, extending the usual picture of the reals as a line. In fact $x + iy$ is obtained by measuring a distance y at right angles to the real line, from the point x. The fact that -1 has no *real* square root is no longer an obstacle in Wallis's picture, because $\sqrt{-1}$ isn't on the real line! It's one unit sideways into the plane. 'Lateral thinking' with a vengeance. But for some reason Wallis's proposal was ignored. In 1797 the Norwegian surveyor Caspar Wessel published a paper showing how to represent complex numbers in the plane. It was in Danish and nobody noticed it either, until a French translation appeared a century later. Meanwhile the

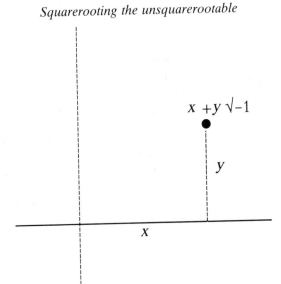

Fig. 17.   Wallis's proposal for visualizing complex numbers.

idea became attributed to Jean-Robert Argand, who wrote about it independently in 1806. By 1811 Gauss was habitually thinking of complex numbers as points in a plane, and he published his ideas in 1831. Finally in 1837 William Rowan Hamilton took the last step, identifying $x + iy$ with its coordinates (x, y) and rewriting the geometric definitions in algebraic form. Thus $(x, 0)$ is identified with the real number x, $(0, 1)$ is i, and we have

$$(x, y) + (u, v) = (x + u, y + v)$$
$$(x, y)(u, v) = (xu - yv, xv + yu).$$

Then

$$i^2 = (0, 1)(0, 1) = (-1, 0) = -1$$

and everything falls sweetly into place. Whenever anyone made a discovery in the mid-1800s it turned out that Gauss had already thought of it, but omitted to tell anyone; and sure enough Gauss had anticipated Hamilton in 1832.

It had now taken some three centuries to reach an answer

that can be explained in three minutes. The answer had been *found* several times but not recognized as such. Moreover, everyone who *was* anyone already *knew* what the answer was. Morris Kline notes that obviously 'many men—Cotes, De Moivre, Euler, Vandermonde—thought of complex numbers as points in the plane' because when they tried to solve $x^n - 1 = 0$ they imagined the solutions as the vertices of a regular n-sided polygon. As if to confirm that everybody knew the answer anyway, we observe the reaction to Hamilton's incisive solution. Did they cheer? Did they roast a hundred oxen? Did they at least buy him a drink? They did not. It was dull, boring, 'old hat'. But even if everyone had known the answer all along, it was only in about 1830 that they started to admit it, even to themselves. What caused the change of heart?

Evaporating philosophy

All the while complex numbers were useless, people thought about them as a problem in philosophy. That meant they had to be invested with some deep and transcendent *meaning*. Thus we find Leibniz making statements like the one that decorates this chapter. For philosophers nothing is better than some obscure but mysterious idea that nobody really cares about and certainly can't test, because then you have plenty of room for clever arguments. Angels on the head of a pin, and so forth. But when something actually becomes useful, most people stop arguing about the philosophy and get on with the job instead. They don't care what the deep philosophical essence of the new gadget is; they just want to churn out as many results as they can in the shortest possible time by taking advantage of it. If you can actually *see* the angels dancing on the head of the pin you stop trying to count them in favour of persuading them to dance on a microchip instead. And that's exactly what happened to complex numbers between about 1825 and 1850. The mathematicians discovered complex analysis—how to do *calculus* with complex numbers. And that turned out to be so powerful that it would all have been dreadfully embarrassing had some ingenious but unwary philosopher proved that complex numbers don't really exist. Philosophical questions ('What *is* this stupid thing?') can sometimes be excuses for *not* getting on with the job of developing an elusive idea. Overnight the complex number concept became so useful that no mathematician in his right mind could possibly ignore it. So the question mutated slightly, into 'What can you *do* with complex numbers?', and the philosophical question ... evaporated. Unmourned, unnoticed, forgotten, buried. There are other cases of this nature in the history of mathematics but perhaps none more clear-cut. As time passes, the cultural world-view changes. What one generation sees as a problem or a solution is not interpreted in the same way by another generation. Today, when the 'real' numbers are seen as no less abstract than many other number systems, complex numbers included, it's hard to grasp how different it all looked to our forebears. We would do well to bear this in mind when we think about the development of mathematics. To view history solely from the viewpoint of the current generation is to court distortion and misinterpretation.

Complex calculus

Unlike the gradual and painful emergence of complex algebra, the development of complex analysis seems to have been a direct result of the mathematician's urge to generalize. It was sought, deliberately, by analogy with real analysis. The theory of limits, differentiation, and power series carry over so easily that they can almost be set as an exercise for a student. The definition of integration carries over easily too; but its properties do not. There is a new, and fruitful, phenomenon. In real analysis, when one integrates a function between two limits, then the limits fully specify the integral. But in the complex case, with the limits represented as points in the plane, there is more freedom. We have to specify a curve linking them, and then 'integrate along the curve'. Since there are lots of possible curves, we need to know to what extent the value of the integral depends on the curve chosen.

In a letter of 1811 to Friedrich Bessel, Gauss (who else?) shows that he knew the theorem that lies at the core of complex analysis. 'I affirm now that the integral has only one value even if taken over different paths provided $f(x)$ does not become infinite in the space enclosed by the two paths. This is a very beautiful theorem whose proof I shall give on a convenient occasion.' But the occasion never arose. Instead, the crucial step of publishing a proof was taken in 1825 by the man who was to occupy centre stage during the first flowering of complex analysis: Augustin-Louis Cauchy. After him it is called Cauchy's Theorem. In Cauchy's hands the basic ideas of complex analysis rapidly emerged. For example, a differentiable complex function has to be very specialized, much more so than in the real case: its real and imaginary parts must satisfy the so-called Cauchy–Riemann equations. Gauss's observation was given a proper proof. If an integral is computed along a path which violates the conditions of the theorem and winds round points where the function becomes infinite, then Cauchy showed how to calculate the integral using the 'theory of residues'. All that's needed is a certain constant, the *residue* of the function at the nasty point, together with a knowledge of how often the path winds round it. The precise route of the path doesn't actually matter at all!

Cauchy's definitions of analytic ideas such as continuity, limits, derivatives, and so on, were not quite those in use today. Full rigour was contributed by Weierstrass, who based the whole theory on power series. The geometric viewpoint was sorely lacking, and Riemann remedied this defect with some far-reaching ideas of his own. In particular the concept known as a *Riemann surface*, which dates from 1851, treats many-valued functions (such as Euler's logarithm) by making the complex plane have multiple layers, on each of which the function is single-valued. The crucial thing turns out to be the topological pattern whereby the layers join up.

From the mid-nineteenth century onwards the progress of complex analysis has been strong and steady with many far-reaching developments. The fundamental ideas of Cauchy remain, now refined and clothed in more topological language. The abstruse invention of complex numbers, once described by our forebears as 'impossible' and 'useless', has become the backbone of mathematics, with practical applications to aerodynamics, fluid mechanics, Quantum Theory, and electrical engineering. Space forbids a detailed survey, but we can capture some of the flavour by looking at two of the most notorious open problems in the area: the Riemann Hypothesis and the Bieberbach Conjecture.

The Riemann Hypothesis

Let $\pi(x)$ denote the number of primes less than x. This is a very irregularly behaved function. But Gauss and others noticed that for large x it is approximately equal to $x/\log x$. This Prime Number Conjecture combined the discrete (primes) and the continuous (logarithms) in an unexpected way, and it is no surprise that it gave rise to some spectacularly original mathematics. The answer turns out to depend on properties of the zeta function

$$\zeta(s) = 1 + 1/2^s + 1/3^s + \ldots$$

mentioned in chapter 4. The relevance to primes was known to Euler, who noticed that

$$1/\zeta(s) = (1 - p_1^{-s})(1 - p_2^{-s}) \ldots$$

where p_n is the nth prime. This formula is not quite as remarkable as it might appear: it's the analyst's way of expressing the existence and uniqueness of prime factorization. But it brings the Big Gun of nineteenth-century mathematics, complex analysis, to bear.

By using the zeta function for real s, Pafnuti Chebyshev in 1852 showed that something close to the Prime Number Conjecture is true: the ratio of $\pi(x)$ and $x/\log x$ lies between 0.922 and 1.105 for large enough x. Bernhard Riemann showed how to define $\zeta(z)$ for all complex z in 1859, by a process called analytic continuation, and tried to prove the conjecture. He realized that the key was to understand the *zeros* of the zeta function, the solutions of $\zeta(z) = 0$. The real zeros are easy to find; they are -2, -4, -6, and so on. Riemann's famous hypothesis, stated quite casually in a short note containing no proofs that sketched out most of the main ideas of the area, is that all the rest are of the form $\frac{1}{2} + iy$, that is, lie on the line 'real part $= \frac{1}{2}$'. It is generally considered to be *the* outstanding problem in mathematics. It has just been shown by J. van de Lune and Herman te Riele that the first 1.5 billion zeros lie on the right line, but for various reasons this is less than compelling evidence. A related conjecture due to Franz Mertens, with similarly strong evidence, was proved *false* in 1984 by Andrew Odlyzko and te Riele.

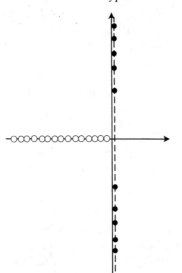

Fig. 18. The first few zeros of the zeta function. Open circles show the 'trivial' zeros at -2, -4, $-6, \ldots$ Black dots show the remaining zeros, whose real part is conjectured to always be $\frac{1}{2}$.

Their proof involved computing the first 2,000 zeros of the zeta function to 100 decimal places. If the Mertens Conjecture had been true, it would have implied the Riemann Hypothesis. The experts weren't entirely surprised that it turned out to be false, because the proof of the Riemann Hypothesis that would result somehow looked 'too easy' to be plausible.

In 1896 Hadamard used the zeta function to prove the prime number theorem. De La Vallée Poussin independently did the same in the same year. If the Riemann Hypothesis is true, the results can be improved by giving a sharp bound on how big the *error* in replacing $\pi(x)$ by $x/\log x$ is. A great many details in the theory of the distribution of the primes depend on the Riemann Hypothesis.

The Zeta function can be greatly generalized. For example, given an algebraic number field, you can use Euler's formula (with prime *ideals* in place of prime numbers) to define its zeta function. There is an appropriate analogue of the Riemann

Hypothesis, which is important throughout algebraic geometry and Diophantine equations. In 1985 there was a flurry of publicity for an announced proof of the Riemann Hypothesis, along these more abstract lines. This announcement was prematune, and the zeta function retains its secrets. Fame, fortune, and many sleepless nights await whoever uncovers them.

The Bieberbach Conjecture

Geometric function theory, as its name suggests, studies the geometry of transformations of the complex plane defined by analytic functions. It has its origins in the discovery that these transformations are *conformal*, that is, do not change the angles between curves. Let the *unit disc* be the set of all complex numbers z whose distance from the origin is less than 1, and let $f(z)$ be a complex function. Then f can be thought of as a way of deforming the disc, moving each z to $f(z)$. The deformed image of the disc may overlap itself, but there is an important class of functions for which it does not. Such a function is said to be *univalent* on the unit disc. At this point the hint of a connection between topological and metric properties arises. The larger the image of the disc, the larger in some sense the function is. The largest univalent function should transform the disc into the largest possible image, namely the entire complex plane. Does this statement make any kind of sense, and if so, what is this largest univalent function? An attractive guess is the Koebe function $z/(1 - z)^2$, because this does have the largest possible image. Its power series expansion is

$$z + 2z^2 + 3z^3 + \ldots + nz^n + \ldots$$

and it occurred to Ludwig Bieberbach in 1916 that its extremality might be reflected in its power series. If so, the nth coefficient in the power series of a univalent function should be no larger than n. In other words, metric extremality follows from geometric extremality. It's an appealing, natural, and simple idea; the sort of thing that it ought to be possible to decide one way or the other without much trouble. In the event, the problem remained unsolved until 1984, when Louis de Branges obtained a miraculous proof.

Until that point the Bieberbach Conjecture was known to be true only for the first six coefficients in the power series. This was proved by Bieberbach himself for the second coefficient, by Karl Löwner in 1923 for the third, by P. R. Garabedian and M. Schiffer in 1955 for the fourth, by R. N. Pederson and M. Ozawa in 1968 for the sixth, and by Pederson and Schiffer in 1972 for the fifth.

De Branges's proof for *all* coefficients is obtained by way of two related conjectures. The first, due to M. S. Robertson in 1936, refers to power series univalent in the unit disc and containing only odd powers of z. The Robertson Conjecture implies the Bieberbach Conjecture. In 1971 I. M. Milin found (but couldn't prove) an inequality which implies the correctness of the Robertson Conjecture. It is this Milin Conjecture that de Branges proves: the Robertson and Bieberbach Conjectures follow immediately.

Löwner's successful attack on the third coefficient is based on a differential equation whose solutions approximate *any* univalent function. It describes how points in the disc 'flow' towards their images under the function, and it is possible to 'push information along the flow'—for example, estimates of the size of power series coefficients. De Branges's proof introduced auxiliary functions to hold the desired information. If these can be shown to have certain properties, then the Milin inequality follows. However, the combinatorics of the sequence of auxiliary functions rapidly becomes unmanageable. Walter Gautschi verified by computer that the desired properties hold for all coefficients up to the 26th (implying the truth of the Bieberbach Conjecture in this range). Later he noticed that the general result follows from a known theorem of Richard Askey and George Gasper about special functions. Askey and Gasper did not have applications to the Bieberbach Conjecture in mind when they proved their theorem: it just seemed an interesting result to them. Most mathematicians think of the field of special functions as an unfashionable and rather pointless dead-end: for this and other reasons that view now seems questionable.

De Branges originally obtained his proof as the climax of a lengthy typescript of the second edition of his book *Square Summable Power Series*. His announcement was greeted with some scepticism. The proof contained errors: it was unclear whether these could be repaired. He had previously claimed a proof of the Bieberbach Conjecture which had turned out to be fallacious, and this damaged his credibility. The conjecture has a history of leading mathematicians astray. As Carl FitzGerald says, 'The Bieberbach Conjecture is not difficult; I have proved it dozens of times.' Acceptance came only after de Branges visited the Steklov Institute in Russia in mid-1984, and his proof was confirmed by the members of the Leningrad Seminar in Geometric Function Theory. The conclusions were written up for the Russian Academy of Sciences and a preprint circulated. At this point mathematicians in the West studied the simplified proof (now translated from the Russian!), made some further simplifications, and agreed that de Branges really had solved the problem. The proof has now been simplified by de Branges, Christian Pommerenke, FitzGerald, and others, and has been made very short and direct: one version of it occupies only five typescript pages.

The exhausted goldmine

We now know that the Bieberbach Conjecture has a short and simple proof, based on standard techniques in geometric function theory. Something very close to the successful assault must have been attempted by dozens, if not hundreds, of mathematicians. Why wasn't the proof found earlier? Perhaps everyone was being stupid? It's easy to criticize with hindsight, but not very helpful. A more likely reason is the subtlety of the problem. The precise approach proves to be very delicate here, and without a great deal of previous work, partial solutions, and 'experimental' results, it is unlikely that anyone would stumble upon the particular combination of conditions needed to make the proof work. Another possibility is that most of these attempts were a bit half-hearted, because nobody seriously thought that the standard approach would do the job. Mathematicians habitually talk of certain methods as being 'worked out', in the mining sense that everything of value that can be obtained from them is already known. It is usually considered to be a waste of time to attack difficult problems by 'exhausted' methods. Yet here, it succeeded.

The answer may be related to the structure of mathematical

proofs. As an analogy, consider a game of chess. Chess openings have been analysed by generations of grandmasters, in the hope of finding a novel line of attack and springing a trap on an opponent who finds himself in unfamiliar territory and under time pressure. But there are so many possible variations in chess that a complete analysis even up to a dozen or so moves is not feasible. There is still room for an orginal line of development, only a little off the beaten track. Nimzovich caused his opponents plenty of headaches by opening with a knight, a move so ridiculous that nobody had bothered to take it seriously. Now there are many more possible 'opening moves' in a mathematical proof than there are in chess. A line of argument may seem to be a bad opening simply because everybody who tried to play the game along that line followed the same pattern of obvious but poor moves, missing a brilliant variation that would crack the game wide open. Certainly it's unwise to write off a plausible line of attack just because nobody has made it work yet.

13

Squaring the unsquarable

'We make no charge for glass balls,' said the
shopman, politely. 'We get them'—he picked
one out of his elbow as he spoke—'free.' He
produced another from the back of his neck,
and laid it beside its predecessor on the counter.

H. G. Wells

One of the more annoying things that mathematicians do is
cast doubt upon things that we imagine we understand perfectly well. For example, we all know what areas and volumes
are. Don't we? I'm going to try to convince you that there's far
more to areas and volumes than we normally imagine. The
tale begins somewhere in the prehistory of the human race,
and culminates in the dramatic solution in 1988 of a problem
posed in 1925 by the mathematical logician Alfred Tarski. He
asked whether it is possible to dissect a circular disc into
finitely many pieces, which can be reassembled to form a
square. Tarski could prove, with some difficulty, that if such a
dissection exists, then the resulting square must have the same
area as the circle; but he wasn't sure whether such a dissection
was possible.

Hang on—isn't it *obvious* that the resulting square must
have the same area? The total area won't be affected by
cutting the circle into pieces, will it? And isn't it *obvious* that
you can't cut a circle up to get a square, because the curved
edges won't fit together properly? No, it's not. Indeed, one of
those two 'obvious' statements is false. The other is true, but
not at all obvious, and its three-dimensional analogue is false.

Squaring the circle

Tarski's problem is often compared to the famous problem of
'squaring the circle', attributed to the ancient Greeks; but

actually it's very different. In the classical formulation, you must construct a square of area equal to that of a circle using ruler and compasses. As we've seen, this was proved impossible by Ferdinand Lindemann in 1882. But Tarski asks

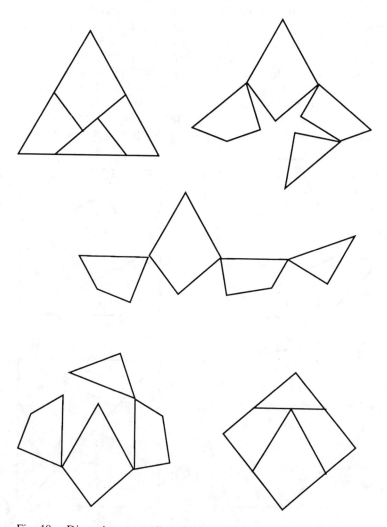

Fig. 19.   Dissecting an equilateral triangle into a square.

for a dissection, not a construction, and you can use any methods you like.

There's a fascinating mathematical toy, which can be made from card but looks best in polished hardwood and brass. It's made by hinging together four oddly-shaped pieces, and attaching handles. If the hinges are closed in one direction, the result is a square; if they are closed the other way, you get an equilateral triangle. This is an ingenious example of a type of puzzle that was very popular towards the end of the nineteenth century, when a lot of people thought that Euclidean geometry was a real fun thing. Such dissection puzzles ask for ways to cut up a given shape or shapes, so that the pieces can be re-assembled to form some other shape or shapes. For example, can you dissect a square into a regular pentagon? A regular

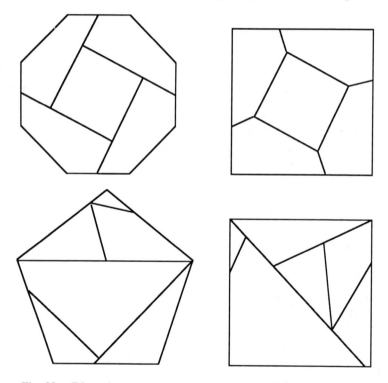

Fig. 20. Dissecting a pentagon and an octagon into a square.

octagon? Both answers are yes (see Fig. 20). What *can* you dissect a square into? For that matter, what can't you?

For simplicity, suppose we restrict ourselves to straight line cuts. Then we always end up with a polygon—a shape with straight edges. It also seems clear that when you dissect a square, its area can't change. Later we'll have to re-examine that belief more critically, but for polygonal dissections it's correct. So, if a square can be dissected to form a polygon, then the polygon must have the same area as the square. Are *all* such polygons possible? The answer is 'yes'. This could have been proved by the ancient Greeks, but as far as we know, it wasn't. It is usually called the Bolyai–Gerwien theorem, because Wolfgang Bolyai raised the question, and P. Gerwien answered it in 1833. However, William Wallace got there earlier: he gave a proof in 1807.

Briefly, the argument runs like this. Begin by dividing the polygon into triangles. Dissect each triangle into a rectangle of the same area. Convert each such rectangle into another rectangle, one of whose sides is that of the desired square (the square root of the area of the original polygon). Finally, stack the resulting rectangles together. The same method proves that any finite set of polygons can be dissected into any other finite set with the same total area.

Area, along with volume, is actually quite a subtle idea, as primary-school teachers know to their cost. It takes a lengthy process of experiment with scissors and paper, and vessels of various sizes, and water all over the floor, before children acquire the concept. In particular, the idea of conservation—that the amount of fluid you have doesn't depend upon the shape of the container that holds it—doesn't come naturally. Most adults can be fooled, by clever packaging, into thinking that they are buying more shampoo than they really are. One difficulty is that areas and volumes are most easily defined for rectangles and boxes: it's not so clear how they work for other shapes, especially ones with curves. We've all been taught that the area of a circle is πr^2; but have you ever wondered *why*? What does π, the ratio of circumference to diameter, have to do with 'how much stuff' there is inside a circle? How does that relate to 'how much stuff' there is inside a square? We saw in chapter 7 that you can't do it with scissors.

Area and volume are such useful concepts that they really do deserve to be put on a sound logical basis. The Bolyai–Gerwein theorem isn't just a recreational curiosity: it's important because it justifies one possible method for defining area. Start by defining the area of a square to be the square of the length of its side, so that for instance a 3 cm square has an area of 9 cm². Then the area of any other polygon is defined to be the area of the square into which it can be dissected. This approach brings out two key properties of area:

Rigidity   The area of a shape stays the same during a rigid motion.

Additivity   If a number of shapes are joined without over-lapping, the resulting area can be found by adding their individual areas.

Actually, there are two versions of additivity. The weaker one, Finite Additivity, applies to a finite number of pieces; the stronger one applies to infinitely many. The technique of integration, basic to the calculus, extends the property of additivity to shapes formed by joining together infinitely many pieces. The total area can then be defined as the sum of the infinite series formed by the areas of the pieces: this is Infinite Additivity. Because infinite series are tricky, this approach requires extreme caution. It is, however, unavoidable if we wish to assign areas to curved regions, such as a circle. A circle cannot be cut into finitely many triangles, but it can be cut into infinitely many. Pursuing a similar line of argument, Archimedes proved that the area of a circle is the same as that of a rectangle whose sides are its radius and half its circumference. Because π is *defined* as the ratio of circumference to diameter, you can now see how it gets in on the act.

Hilbert's hit-list

David Hilbert, one of the greatest mathematicians of all time, wondered whether an equally 'clean' approach to volume might be possible. Mathematicians already knew what volume *was*, of course: in particular they could calculate the volume of any pyramid (one third the height times the area of the base). But again they wanted sound logical foundations. Hilbert asked whether any polyhedron (solid bounded by flat surfaces) can always be dissected into a finite number of pieces, and reassembled to form a cube of the same volume. In 1900, at the International Congress of Mathematicians in Paris, he listed the question among twenty-three major unsolved problems.

Unlike the other twenty-two, it didn't survive very long. In the following year Max Dehn, one of the founders of topology, found the answer. Surprisingly, it is 'no'. Dehn's proof uses the angles of the solid to define what is now called the *Dehn invariant*, a number that, like volume, remains unchanged under dissection and reassembly. However, solids of the same volume can have different Dehn invariants. For example, this happens for a cube and a regular tetrahedron. So you can't dissect a tetrahedron into a cube of equal volume. The Dehn invariant is the only new obstacle to dissectability in three dimensions, and the correct analogue of the Bolyai–Gerwien theorem is that polyhedra can be dissected into each other if and only if they have the same volume *and the same Dehn invariant*.

Though differing in this respect, area and volume share the basic features of rigidity and infinite additivity. A general theory of such concepts was developed by Henri Lebesgue, who called them 'measures', and devised the great grandaddy of all such concepts, Lebesgue measure. Along the way, it transpired that sufficiently messy and complicated sets may not possess well-defined areas or volumes at all. Only the 'measurable' sets do. These include all of the familar shapes from geometry, and a wide range of much more bizarre sets as well; but not everything. Measure theory is important for a variety of reasons: in particular, it is the foundation for probability theory.

Double your money

Tarski belonged to a group of Polish mathematicians who frequented the celebrated Scottish Café in Lvov. Another member was Stefan Banach. All sorts of curious ideas came out of bull sessions in the Scottish Café. Among them is a theorem so ridiculous that it is almost unbelievable, known as the Banach–Tarski Paradox. It dates from 1924, and states that it is possible to dissect a solid sphere into six pieces, which can be reassembled, by rigid motions, to form two solid spheres each the same size as the original.

But what about the volume? It doubles. Surely that's impossible? The trick is that the pieces are so complicated that *they don't have volumes*. The total volume can *change*. Because the pieces are so complicated, with arbitrarily fine detail, you can't actually carry out this dissection on a lump of physical matter. A good job too, it would ruin the gold market.

Once we start thinking about very complicated pieces, we have to be more careful. For instance, the 'pieces' need not actually come as connected lumps. Each 'piece' might consist of many disconnected components—possibly infinitely many. When we apply a rigid motion to such a set, we must not only keep each component the same shape: we must also preserve the mutual relationship of the parts. And it's not just the notion of rigid motion that needs to be made precise. When we dissected a square into an equilateral traingle, for example, we didn't worry about the edges of the pieces. Do they abut or overlap? If a point is used in one edge, does that exclude it from the adjacent edge? The question wasn't raised before because it would have been distracting; and also because the edges of polygons have zero area and hence don't really matter. But with highly complicated sets, we need to be more careful: *every single point* must occur in precisely one of the pieces. Banach proved the Bolyai–Gerwein theorem is still true for this more careful definition of 'dissection'. He did so by inventing a method for 'losing' unwanted edges.

The spirit (but not the details) of how the Banach–Tarski Paradox works can be understood by thinking about a dictionary rather than a sphere. This is an idealized mathematician's dictionary, the *Hyperwebster*, which contains all possible words—sense or nonsense—that can be formed from the 26 letters of the English alphabet. They are arranged in alphabetical order. It begins with the words A, AA, AAA, AAAA, . . . and only infinitely many words later does it get round to AB. Nevertheless, every word, including AARDVARK, BANACH, TARSKI, or ZYMOLOGY, finds its place. I'll show you how to dissect a Hyperwebster into 26 copies of itself, each maintaining the correct alphabetical order—with a spare alphabet thrown in.

The first of the 26 copies, 'volume A', consists of all words that begin with A, apart from A itself. The second, 'volume B', consists of all words that begin with B, apart from B itself; and so on. Let's think about volume B. Like gentlemen songsters out on a spree, it begins BA, BAA, BAAA, and so on. Indeed, it contains every word in the entire Hyperwebster exactly once, except that a B has been stuck on the front of every one. BAARDVARK, BBANACH, BTARSKI, and BZYMOLOGY are in volume B. Moroever, they are in the same order as AARDVARK, BANACH, TARSKI, and ZYMOLOGY.

The same goes for volume A and volumes C–Z. Each is a perfect copy of the entire Hyperwebster, with an extra letter stuck on the front of every word. Conversely, every word in the Hyperwebster, apart from those containing just a single letter, appears in precisely one of these 26 volumes. AARDVARK, for instance, is in volume A, in the position reserved for ARDVARK in the Hyperwebster itself. BANACH is in volume B, and its position corresponds to that of ANACH in the original Hyperwebster. The cartoonist's snooze symbol ZZZZZ is in volume Z, in the position corresponding to the slightly shorter snooze ZZZZ.

In short, one Hyperwebster can be cut up, and rearranged, without altering the orders of the words, to form 26 identical Hyperwebsters plus a spare alphabet. Order-preserving dissections of Hyperwebsters are not 'volume'-preserving. For 'Hyperwebster' now read 'sphere', for 'word' read 'point', and for 'without altering the order of' read 'without altering the distances between', and you've got the Banach–Tarski Paradox. In fact the analogy is closer than it might appear, because the Banach–Tarski Paradox is proved by using sequences of rigid motions in much the same way as I have used sequences of letters. The possibility of doing this was discovered by Felix Hausdorff, who showed that there exist 'independent' rotations, for which every different sequence of combinations leads to a distinct result—just as every different sequence of letters leads to a distinct word in the Hyperwebster. In spirit, the Banach–Tarski Paradox is just the Hyperwebster Paradox wrapped round a sphere.

Banach and Tarski actually proved something much stronger. Take any two sets in space, subject to two conditions: they don't extend to infinity, and they each contain a solid sphere, which can be as small or as large as you wish. Then you can dissect one into the other. Squaring the Circle may be tough, but Banach and Tarski could Cube the Sphere—getting any size of cube, to boot. They could dissect a football into a statue of Lady Godiva, a scale model of a rabbit three light years high, or a pinhead in the shape of the Taj Mahal.

Plane speaking

The Banach–Tarski Paradox doesn't arise in two dimensions. You can't change areas in the plane by dissection. (You *can* change areas on the surface of a sphere.) The basic reason is that in the plane, the result of performing two rotations in succession is independent of the order. Analogously, in the corresponding *Dyslexicon*, the order of the letters doesn't matter: SPOTTER means the same as POTTERS. Now the dissection trick goes haywire: if we put SPOTTER in volume s, corresponding to POTTER in the original, then volume P fails to contain anything corresponding to OTTERS. So the dictionary approach certainly won't work. Tarski proved that nothing else would either. Banach had proved the existence, in the plane, of what is now called a Banach measure. This is a concept of area, corresponding to the usual one for those sets that *have* areas in the usual sense, and defined for every set whatsoever, even those that *don't* have an area, which is rigid and *finitely* additive. It follows easily that no two-dimensional Banach–Tarski Paradox is possible.

  That's the point Tarski had reached when he posed his circle-squaring problem. *If* a circular disc can be cut into a finite number of pieces, which can be reassembled to form a square, *then* the square must have the 'correct' area. However, it wasn't clear to him whether such a process is possible.

  The problem proved difficult, so mathematicians concentrated on special cases: restricting the nature of the pieces, or the type of rigid motion allowed. In 1963 L. Dubins, M. Hirsch, and J. Karush proved that you can't cut up a circle along continuous curves and reassemble the pieces to form a square. You can't market a jigsaw puzzle whose pieces can be put together one way to form a circle and another way to form a square. If the circle can be squared by dissection, then the 'pieces' have to be far more complex than anything that can be cut by the finest jig-saw.

In 1988, at a conference in Capri, an American mathematician named Richard Gardner gave a lecture. In it he proved that, in any number of dimensions and not just two, there can be no solution to a version of Tarski's circle-squaring problem in which a limited system of rigid motions is permitted (selected from any 'discrete group'). He conjectured that the same is true if a rather broader system of rigid motions is allowed ('amenable group'). In the audience was Miklos Laczkovich, a Hungarian mathematician from Eötvös Lorand University in Budapest. Laczkovich had been making a careful study of dissection problems in *one* dimension, and during Gardner's talk he realized that his conjecture is *false* in the one-dimensional setting.

Marriage made in heaven

The discovery triggered a whole series of new ideas on Tarski's circle-squaring problem itself, and within two months Laczkovich had a complete solution. You *can* square the circle by finite dissection! Even more suprisingly, the pieces do not even have to be rotated. The same goes for any shape whose boundary is composed of smoothly curving pieces, such as an ellipse, or a crescent, or the outline of an amoeba.

Laczkovich's method requires about 10^{50} pieces, so it's not something you can draw. The proof occupies about forty pages, and involves a whole series of novel ideas: naturally, it gets rather technical. One trick worth mentioning is a version of the Marriage Theorem. A (rather old-fashioned) dating agency has a rule that clients may date only clients of the opposite sex to whom they have been properly introduced. Under what conditions can all clients secure a date simultaneously? Obviously the numbers of men and women have to be equal, but that's not enough—for instance, some men may not have been introduced to any women at all. Even if each man has been introduced to at least one woman, there can still be problems: two hundred men trying to date one woman, who has been introduced to them all. The more you think about it, the trickier it gets; but there's a very elegant solution. According to the Marriage Theorem, a complete set of dates is

Squaring the unsquarable

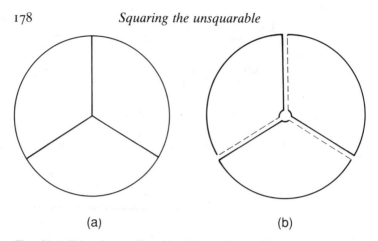

(a)   (b)

Fig. 21.   Trisecting a disc (a) with overlaps, (b) minus its central point.

possible if and only if every subset of men collectively has been introduced to at least as many women, and conversely. In Laczkovich's proof the 'men' are possible candidates for pieces of the square, the 'women' are candidates for pieces of the circle, and 'being introduced' is the same as 'related by a rigid motion'. The Marriage Theorem lets Laczkovich 'marry' a suitable collection of pieces from the square to a corresponding collection from the circle. Much of the remainder of the proof tackles the construction of those candidate sets.

Despite Laczkovich's brilliant breakthrough, the last word has not been said. Here's a (deceptively) simple example of the questions that remain unanswered. *Can a circular disc be trisected*? That is, can you cut a circular disc into three pieces which, apart from rigid motions, are *identical*? It's easy to do this if overlaps at edges are allowed, or for a disc with its central point missing. The central difficulty is to deal with that one tricky point.

14

Strumpet Fortune

Out, out thou strumpet, Fortune! All you gods,
In general synod, take away her power,
Break all the spokes and fellies from her wheel,
And bowl the round knave down the hill of heaven,
As low as to the fiends.

William Shakespeare

Not a nice thing to say about a lady, but a time-honoured
judgement on the character of Lady Luck. But does the Lady
deserve her scarlet reputation?

At the end of the eighteenth century the 'system of the
world' was a clockwork-like machine, set going by the Hand of
Creation but thereafter left to impose its own precision on
the universe. The Laws of Nature were sharp-edged and predictable; nothing was left to chance, not a leaf fell nor wavelet
rippled without exact and meticulous cause. Chance confined
herself to sleazy pursuits, such as gambling. In the twentieth
century, we find ourselves in a world ruled by chance, in which
the very stuff of Nature is a random cloud of probabilities.
Clockwork has given place to a Cosmic Lottery. But randomness is no longer synonymous with patternless chaos
and capriciousness. Chance too has its patterns. Demurely
obedient the Lady is not, but for all that she is no wanton. She
creates her own precedents, follows her own paradigm; but
there is an inner order to her conduct. And that inner order
has given rise to a new kind of mathematics: the theory of
probabilities.

The fair coin

It is not inappropriate that the earliest writings on probabilities
were by Girolamo Cardano, the Gambling Scholar—and that
they had to do with those most practical of human activities,

games of chance. Probability theory, and its applied arm, statistics, has had a sometimes uneasy relationship with gambling ever since. For example life insurance: you bet that you will die, the insurance company that you will not. In order to win, you have to lose! The particular problem that interested Cardano was how to divide the stakes for a game of dice that is interrupted. The same question was asked of Blaise Pascal by his friend the Chevalier de Meré in 1654. A player is given eight tries to throw a one with a die, but the police raid the joint after three throws have been completed: how should the stakes be shared out after the hue and cry has died down? The question prompted Pascal to write to Fermat, and their thoughts were published by Huygens in 1657 in the first book devoted to probability theory: *On Reasoning in Games of Dice*. The investigation led Pascal to his famous arithmetical triangle

in which each number is the sum of that directly above it and that above and to the left. For example, consider three tosses of a coin. There are eight possible outcomes in all, which I'll arrange like this:

HHH	HHT	HTT	TTT
	HTH	THT	
	THH	TTH	

Of the eight series of tosses, three heads occur in one case, two heads in three, one head in three, and no heads in one. The sequence of numbers is 1, 3, 3, 1, just like the fourth row of Pascal's triangle; and their sum is 8. The chances of throwing exactly one head in three trials are 'three out of eight', or to put it another way, the probability of throwing exactly one head in three trials is 3/8. Assuming, of course, that all possibilities are 'equally likely', or that the coin is *fair*.

The arrival of probability theory as a subject in its own right may be traced to Pierre-Simon de Laplace's *Analytic Theory of*

Probabilities of 1812. Here he defines the probability of a particular event as the ratio of the number of ways in which it can happen, compared to the total number of possible outcomes—again on the assumption that all outcomes are equally likely. What do we mean by 'equally likely'? To answer that the probabilities are all the same is to invoke circular reasoning! This foundational question caused a lot of trouble in the early days of probability theory: we shall see later how it has been resolved. Laplace's definition reduces questions of probability to combinatorial counting problems. For example, to find the chances of getting a full house (three cards of one value and two of another, say three kings and two queens) in a game of 5-card stud, we work out

(1) The total number of different poker hands (2598960)
(2) The number of full houses (3744)

and form their ratio

$$3744/2598960 = 0.00144.$$

Problems with the infinite

This approach limits probability to discrete—indeed finite—systems. But many practical problems involve infinitely many possible outcomes. For example, throw three darts at a dartboard: what is the probability of getting three treble-20s? Modelling the board as a circular disc and the position of the dart as a point, there are infinitely many possible outcomes, of which infinitely many are favourable. Laplace's recipe leads to ∞/∞ which doesn't get us very far. It is also not clear what we should mean by 'equally likely' here. For each individual point the probability of hitting it is, again by Laplace, $1/\infty$, which is 0. The chance of hitting any given point is exactly zero. So all points are equally likely. But the chances of hitting some set of points should be the sum, over that set, of the chances of hitting individual points. The sum of lots of 0s is 0. How can this be reconciled with the fact that, on any throw, *some* point is definitely hit?

The simplest approach is to assume that the chance of hitting a given *region* of the board is proportional to its *area*. Laplace's ratios of numbers are replaced by ratios of areas.

For simplicity choose units making the total area 1. The area
of a single point is zero, and so are the chances of hitting it;
but the area of the whole board is 1, and we are therefore
bound to hit *something*. The area of the treble-20 is roughly
0.0005; the probability of three-in-a-bed is the cube of this,
about one chance in eight billion. Of course a good darts
player does better, since he does *not* distribute darts over the
board at random, but concentrates them near the target zone.
The area should be appropriately 'weighted' to take this into
account.

During the eighteenth and nineteenth centuries mathematicians grappled with such problems, but failed to find
a really precise formulation of the general notion of probability. Laplace said hopefully that 'at bottom, the theory of
probabilities is only common sense, reduced to calculation',
but unfortunately there are situations in which common
sense, applied in equally reasonable ways, gives contradictory
answers. As Morgan Crofton put it in 1868, 'It is true that in a
few cases differences of opinion have arisen as to the principles, and discordant results have been arrived at', but added
'all feel that this arises, not from any inherent ambiguity in the
subject matter, but from the weakness of the instrument
employed.'

A stronger instrument

The next significant step apparently had nothing to do with
probabilities, but arose in the calculation of lengths, areas, and
volumes. The way to do this is to use the integral calculus. The
area under a curve is the integral of the function that defines
it; volumes and higher-dimensional analogues thereof are
obtained by multiple integrals. There are technical difficulties
in the manipulation of multiple integrals that led mathematicians to seek a more general notion of the basic concepts
of length, area, and volume. The hope was that the greater
generality would resolve the technical snags.

The key concept is length. If a and b are two points on the
real line, then the length of the interval joining them is $b - a$.
Given a set of points on the real line built up from a finite
number of intervals, then it is reasonable to define its length to
be the sum of the lengths of its component intervals. But more

complicated sets can be envisaged. Is there a way to define the 'total length' of *any* set of real numbers? In 1882 du Bois-Reymond defined the *content* of a set by 'approximation from the outside' by finite unions of intervals. Peano improved the idea in 1887 by approximating from inside as well. For example, to find the area of a region in the plane, he calculated the *outer content* as the limit of the areas of polygons containing the set, and the *inner content* as the limit of the areas of polygons contained within the set. If these are equal, their common value is defined to be the content, or area, of the set. (If not, the set is considered not to have an area.) This is a direct generalization of Archimedes's method for finding the area of a circle. Put a polygon outside, another inside, then increase the numbers of sides of the polygons so that the area between them shrinks to zero. The advantage of Peano's concept is that it is *additive*: to find the area of a set that splits up into finitely many pieces, you just add up the individual areas.

The definitive version of the theory was devised by Henri Lebesgue in 1902. In Lebesgue's theory, the notion of content is replaced by *measure*. The approximating sets are not unions of finitely many intervals, but of an infinite sequence of intervals. The total length is still defined provided the infinite series formed by the individual lengths converges to a well-defined sum. And Lebesgue's measure is not just additive: it is *countably additive*. If a set is built up from a sequence of disjoint pieces, then its measure is the sum of the measures of the pieces (interpreted as an infinite series). With this definition, a great many complicated sets can be shown to possess lengths, areas, and so on. For example, the set of rational numbers between 0 and 1 has measure 0; the set of irrationals between 0 and 1 has measure 1. As far as measure theory goes, there are a lot more irrationals than rationals (a judgement with which Cantor, for other reasons, would have agreed). But even with this much broader definition, there remain sets which do not have measures (at least in the usual formulations of set theory). Despite this, the range of *measurable* sets is enormous.

Lebesgue's ideas (arrived at independently by William Young) led to a greatly improved theory of integration, and have become a cornerstone of analysis. But they also solve the problem of continuous probability theory.

Kolmogorov's axioms

As on many previous occasions, the answer was to stop worrying about what probability 'really' is, and instead to write down a list of properties that it ought to have. In other words, to axiomatize the concept. The final step was taken in about 1930 by Andrei Kolmogorov, using the idea of a *sample space* due to Richard von Mises: the set of possible *outcomes* of an experiment. For example, if a coin is tossed once, the sample space is the set {Heads, Tails}. If a dart is thrown, it is the set of points on the dartboard. If three coins are tossed, it is the set of eight possible sequences of heads and tails. Now comes the tricky step. Define an *event* to be not just a single outcome, but a *set* of outcomes. For example, when tossing three coins, the event 'two heads and one tail' corresponds to the set {HHT, HTH, THH} of outcomes. In the case of a dart being thrown, the event 'treble-20' corresponds to a particular region of the dartboard, in the treble ring and the 20 sector; and this region is again a set (this time infinite) of points of the sample space.

We now select a suitable set of events (I'll explain what 'suitable' means in a moment) and define a *probability function p*, which assigns to each event E in the set a number $p(E)$. This should have the following properties:

(1) $p(E)$ always lies between 0 and 1.
(2) If E is the empty event 'nothing happens' then $p(E) = 0$.
(3) If E is the universal event 'some possible outcome happens' then $p(E) = 1$.
(4) p is *countably additive* in Lebesgue's sense.

'Suitable' just means that given any sequence of events E for which $p(E)$ is defined, they can be lumped together into a single event F (their union) for which $p(F)$ is also defined. The set of events considered must be 'closed under countable unions'.

How does this help?

For finite probability everything is very straightforward. The set of events to be considered contains *all* sets of outcomes. Assign probabilities to individual outcomes so that the total is 1; then the probability $p(E)$ of an event (set of outcomes) E is the sum of the probabilities of its members. For example in the tossing of three coins, each of the eight possible sequences is *assigned* probability 1/8; and this is the *definition* of a fair coin. The meaning of the phrase 'equally likely' doesn't enter into the formulation; but of course it provides the motivation for the values chosen. For the dartboard example, we must be more careful. The events considered are now those *regions* of the dartboard that possess an area, in the sense of measure theory. The probability $p(E)$ is the proportion of the total area filled up by E. Individual points have area zero, hence probability zero; but the treble-20 region R has a definite non-zero area, and hence a non-zero probability. Now R can be broken up into infinitely many 1-point subsets, each of probability zero. So why does not property (4) imply that $p(R)$ is the sum of these, hence itself zero? The answer is that R is made up from an *uncountable* infinity of points, so we cannot appeal to (4), which asserts only countable additivity. It is on this delicate distinction that the formulation hinges, and the apparent paradoxes are resolved.

In an application of probability theory, it is up to the end user to decide for himself what the sample space, the set of events, and the probability function should be. Only then does the machinery of probability theory swing into action. But this is entirely right and proper. The laws of mechanics, for example, tell us how to analyse the motion of the Solar System; but they do not of themselves prescribe the masses or shapes of the planets, or their initial positions in orbit. This modelling step must be made first, before the mathematics can be used: it is not part of the mathematics.

The multidimensional drunkard

Among the things probabilists study are *stochastic processes*. Here a system of some kind sits in some particular state, and at each instant the state changes randomly according to some specified set of probabilities. One possible application is to the motion of a molecule under random collisions with other molecules; that is, the diffusion of a chemical through some medium. Now it's possible to treat diffusion effects in the time-honoured way, and develop a continuum-mechanical model based on partial differential equations. We may now ask whether this is in agreement with the predictions of a molecular theory. The mathematics for molecular diffusion is complicated, and it makes sense to start with simple models and work our way up to something more realistic.

The simplest such process is the one-dimensional *random walk*. Imagine a line, marked off in units. At time 0 a particle starts at position 0. At time 1 a coin is tossed: if the coin comes up heads the particle moves one unit to the right; if tails, one unit to the left. At time 2 this process is repeated, and so on. Assuming the coin to be fair, the *transition probabilities*, for motion right or left, are $\frac{1}{2}$ at each stage. We ask what the long-term behaviour of this system looks like. Will the particle wander off along the line, or hover about near the origin, or what? Since it has equal chances of going either way, it doesn't seem especially likely that it will wander away for good; on the other hand, there's little to prevent it drifting quite a way along the line, because every so often a run of unusually many heads or tails will occur. In other words, we expect its position to fluctuate in a pretty irregular fashion, not moving too fast on average, but with a fair chance of eventually getting to pretty much any point on the line. And indeed that's what happens. Given *any* point, however far away from the origin, the probability that the particle will eventually reach that point is 1. Or, putting it another way, the chance that it *never* reaches the chosen point is zero. Indeed, with probability 1 it returns infinitely often to the point. The motion is smeared into 'uniform' fluctuations over the whole line.

So, if you are lost in a 1-dimensional desert, and go East or West by repeatedly tossing a coin, you will eventually reach any point you wish. But there's a price to pay: it takes a very long time. In fact the time is so long that the *average* time between successive returns to the same point is infinite. If you perform a similar random walk, but now move in the plane, with probabilities $\frac{1}{4}$ of going East, West, North, or South, the results are very similar. If you are lost in a 2-dimensional desert, you still expect to reach every point eventually by moving at random in each of the four directions. What about three dimensions? Now you can go up and down as well, with all transition probabilities being $\frac{1}{6}$. In 1921 George Pólya proved that the results are very different: there is a non-zero probability (about 0.65) that you *never* reach your chosen position, however long you walk.

These examples are obviously very special, although they already make the point that the dimension of the space in which the walk takes place has a strong influence. A great many generalizations suggest themselves. There is no reason to make all of the transition probabilities the same. How do the results depend on the values chosen? What happens if we allow moves in 'diagonal' directions? Along a specified set of directions? The random walks discussed above are *discrete* in that the positions are specified by *integer* coordinates and the distance moved at each stage is the same (namely 1). What about *continuous* random walks? What if the discrete time-steps are replaced by a continuous motion?

The molecular speculator

In 1827 the botanist Robert Brown was staring down his microscope at a drop of fluid. There were tiny particles in the fluid, and he noticed that they weren't sitting still: they were jumping about in a very erratic fashion. The particles weren't alive, moving under their own volition; and the fluid was absolutely still. What was the cause? Brown suggested that this movement is evidence for the molecular nature of matter—which at the time was itself a highly speculative theory. The fluid is composed of tiny molecules, whizzing about at high speed and colliding randomly with each other. When the molecules bump into a particle, they give it a random push.

It is possible to model Brownian motion by a 'continuous' random walk. This is defined by taking a discrete version, with small time-steps and small individual movements in many directions, and then passing to the limit. William Feller describes what this achieves:

In the limit the process will occur as a continuous motion. The point of interest is that in passing to the limit our formulas remain meaningful and agree with physically significant formulas of diffusion theory which can be derived under much more general conditions by more streamlined methods. This explains why the random-walk model, despite its crudeness, describes diffusion processes reasonably well.

What we have is a direct mathematical connection between the motions of individual molecules, and the bulk properties of the fluid they make up. The statistical features of molecular motion, and the continuum mechanics of diffusion, agree.

The first person to discover the connection between random walks and diffusion was Louis Bachelier, in a doctoral dissertation of 1900. The examiners, damning it with faint praise, gave it a *mention honorable* at a time when only a *mention très honorable* was considered worth having. Bachelier wasn't studying Brownian motion, however: he had his eye on something closer to the gambling origins of probability theory—the random fluctuations of the Paris stock-market. It's perhaps no surprise that a failed thesis on stocks and shares attracted little attention from theoretical physicists. In 1905 Albert Einstein laid what everyone thought were the foundations of the mathematical theory of Brownian motion; and Norbert Wiener developed these extensively. Only much later was it discovered that Bachelier had anticipated many of their main ideas!

Random walks are still important in mathematical physics, and are now appearing in quantum theory. Among the unsolved problems is the behaviour of *self-avoiding* random walks, in which the particle is not permitted to visit the same point twice. Meanwhile the mathematics of probability theory, in innumerable forms from statistical mechanics to ergodic theory to stochastic differential equations, has diffused throughout the whole of the sciences.

15

The mathematics of nature

And new Philosophy calls all in doubt,
The Element of fire is quite put out,
The Sun is lost, and th'earth, and no mans wit
Can well direct him, where to looke for it.

John Donne

In 1642 in the village of Woolsthorpe, a widow who ran the family farm was blessed with a son. Born prematurely, he was not expected to live beyond a week, but he survived. He went to ordinary schools and was an ordinary little boy, save perhaps for a knack for making gadgets. A stay at Grantham is described thus by Richard Westfall:

While the other boys played their games, he made things from wood, and not just doll furniture for the girls, but also and especially models. A windmill was built north of Grantham while he was there. Only the schoolboy inspected it so closely that he could build a model of it, as good a piece of workmanship as the original and one which worked when he set it on the roof. He went the original one better. He equipped his model with a treadmill run by a mouse which was urged on either by tugs on a string tied to its tail or by corn placed above it to the front. He called the mouse his miller. He made a little vehicle for himself, a four-wheeled cart run by a crank which he turned as he sat in it. He made a lantern of 'crimpled paper' to light his way to school on dark winter mornings, which he could simply fold up and put in his pocket for the day. The lantern had other possibilities; attached to the tail of a kite at night, it 'wonderfully affrighted all the neighbouring inhabitants for some time, and caused not a little discourse on market days, among the country people, when over their mugs of ale.'

Growing older, the young man studied at Trinity College, Cambridge, making no special impact. In 1665 Europe was under the scourge of a disease transmitted to man by rat-fleas and causing swelling of the lymph glands: the bubonic plague.

The young man returned to the safety of his home village. In rural solitude he almost single-handedly created mechanics, optics, and the calculus. His name was Isaac Newton.

Why did the dormant genius awaken so explosively? The best explanation we have is Newton's own: 'In those days I was in the prime of my life for invention, & minded Mathematicks & Science more than at any time since.' He published none of this work. According to De Morgan the reason was that 'a morbid fear of opposition from others ruled his whole life'. In *The Sotweed Factor* John Barth's fictional Newton was 'so diffident about his talents 'twas with great reluctance he allowed aught of his discoveries to be printed; yet so vain, the slightest suggestion that someone had ante-dated him would drive him near mad with rage and jealousy. Impossible, splendid fellow!' And so it may well have been. Finally in 1672 Newton's work on optics appeared in print, and he promptly found his fears justified, running into a united wall of opposition from scientists such as Robert Hooke and Christiaan Huygens.

This was hardly encouraging. But, after much urging by Edmund Halley, Newton did eventually publish his theories of mechanics and gravitation, the *Philosophiae Naturalis Principia Mathematica*; probably *the* most significant book in history, for it explained the 'System of the World'. Unlike most attempts at this, it succeeded. It was a difficult book, and deliberately so; lesser men than Newton have also deflected criticism with a competent snow-job. As far as proofs go, the *Principia* is relentlessly geometrical, in the classical mould. But there is a strangeness to Newton's geometry. A competent mathematician can follow it and verify its correctness; but never in a month of Sundays could he answer the only import-ant question—how the devil did Newton think of it? Many years later Gauss, apropos of proofs, remarked 'When one has constructed a fine building the scaffolding should no longer be visible', and Newton was apparently of like mind. Jacobi called Gauss 'the fox of mathematics' because he erased his tracks in the sand with his tail. The tracks Newton was busily erasing were those of the calculus. He had in his possession a devastatingly powerful weapon. He made sure it stayed a concealed weapon; either to make his methods seem more familiar to his contemporaries, or to avoid more controversy, or possibly because of his deep admiration for classical Greek geometry. His successors, less reticent and able to spot a worthwhile dodge when they saw it, adopted the calculus explicitly. The result was a new paradigm for the description of Nature: *differential equations*.

Vive la differential

Galileo Galilei had investigated the motion of bodies and derived mathematical laws from his observations, saying prophetically:

It has been pointed out that missiles and projectiles describe a curved path of some sort; but no one has pointed out the fact that this path is a parabola. But this and other facts, not few in number or less worth knowing, I have succeeded in proving; and what I consider more important, there have been opened up to this vast and excellent science, of which my work is merely the beginning, ways and means by which other minds more acute than mine will explore its remote corners.

Newton was acute enough to seek general laws governing all kinds of motions, and this problem led him to invent calculus. Here's a modern summary. Let x be a quantity depending on time. Then it has an instantaneous rate of change, or *derivative*, written x'. For example if x is position, then x' is velocity. Repeating the process we get x'', which here is acceleration; and so on. Newton called these objects *fluxions*, and the original quantities they were derived from he called *fluents*.

Newton's basic Law of Motion states that the *acceleration* of a body is proportional to the force acting on it, that is,

$$mx'' = F$$

where m is its mass, x its position, and F the force acting. This is a *differential equation* because it specifies not x, which we want to know, but a derivative (here the second). Its solution depends on how F varies with time, and is by no means evident. The variety of possible differential equations is endless; the number of natural phenomena they can describe is scarcely less. These new types of equations demand new methods of solution. Newton favoured one above all: power series. He assumes that the variation of x with time t is of the form

$$x = a_0 + a_1t + a_2t^2 + \ldots,$$

then substitutes this in the equation, and solves it in turn for $a_0, a_1, \ldots$ Others, such as Leibniz, preferred to find solutions in 'closed form' that is, as a formula in t. Neither approach is really satisfactory. 'Closed form' solutions seldom exist, though they're useful when they do, and many an equation has been torn limb from limb until some mutilated corpse has a solution in closed form—to what purpose is debatable. Power series solutions often only work for small values of the variables, and sometimes not even then—though an important idea, the theory of asymptotic series, makes sense of some series even when they don't converge.

Sun-god and crystal spheres

The immediate impact of Newton's work was greatest in that time-honoured subject, celestial mechanics: the motion of the heavenly bodies. The astronomers of ancient Babylon made many accurate observations and understood the heavens well enough in empirical terms to predict eclipses. Their theoretical cosmology was less satisfactory, being compounded of two earlier ones, those of Eridu and Nippur. In Eridu cosmology water is the origin of all things. The world arises from the sea and is encircled by it; beyond the ocean-stream the sun-god pastures his cattle. In that of Nippur the world is a mountain. The Babylonians held that the heavens were a solid vault with its foundations on the ocean. Above are the upper waters; above these live the gods, and the 'sun-illuminated house' from which the Sun emerges every morning through one door and to which it returns every evening through another. Under the mountain lies the abode of the dead. Old Testament cosmology echoes these ideas.

The Egyptians held that the Earth is a long box, running north-south (like the Nile). The sky is the ceiling; stars hang on lamps by cords or are carried by deities. During the day they are switched off. The ceiling is supported by four mountains at the cardinal points. On a ledge below their peaks a river runs round the Earth, the Nile being one of its branches.

The sun-god Ra is carried round this river in a boat, and when he is attacked by a serpent, there is an eclipse. The Nile was the source of all life in Egypt, and it consequently loomed large in Egyptian cosmology.

The Ionian school of Greek philosophers made no significant advance on this picture. Anaximander held that the Earth is flat; Anaximenes that the celestial vault is solid, with stars attached like nails. Thales did predict an eclipse in 585 BC, but he probably got the method from the Egyptians or Chaldeans. Parmenides, in about 500 BC, realized that the Earth is a sphere, but only the Pythagoreans believed him, and they probably for the wrong reasons. He stated that the Earth is immobile because there is no good reason why it should fall one way or another. The Pythagoreans elaborated the theory, embellishing it with their characteristic number mysticism. There are ten heavenly bodies, they said, since $10 = 1 + 2 + 3 + 4$ is the source of all number. These are the Earth, Moon, Sun, Venus, Mercury, Mars, Jupiter, Saturn, the Starry Sphere, and Antichthon (counter-Earth). All encircle a celestial fire, with Antichthon perpetually hidden from us by the Earth (which also conveniently hides the fire). The Earth is a sphere; the Moon does not rotate. The Sun takes a year to encircle the central fires, the Earth only 24 hours: this explains day and night. The planets emit harmonious tones (the 'music of the spheres') which we don't hear because we are so used to them.

According to Plato, the Earth is round and lies at the centre of the universe, with everything else turning about it on a series of hollow spheres, rotating on their axes. Eudoxus modified this model to incorporate reversals of direction that had been observed in the motion of planets, and the inclination of the planetary orbits. He mounted the poles of one sphere on another, the poles of this on the next, and so on, using 27 spheres in all. Kalippus improved the fit with observation by extending the system to 34 spheres. Aristotle accepted Eudoxus's spheres but rejected celestial harmony. He treated the spheres as actually existing, and hence had to add new ones to prevent the motion of one set being transmitted to the next, getting 56 in all. Herakleides (370 BC) held that the Earth rotates, the stars are fixed, Venus moves round the Sun, not the Earth, and so does Mercury. Aristarchus (281 BC) suggested that the Earth goes round the Sun. When it was observed that the Moon varies in size, hence cannot go round the Earth in a circle, a small eccentric circle was added. This still wasn't satisfactory, so Apollonius (230 BC) used *epicycles*. The Earth goes round a circle whose centre goes round another circle. Claudius Ptolemaeus (Ptolemy) lived in Alexandria in AD 100–160. He proved that anything you can do with an eccentric can be done using an epicycle, so eccentrics lapsed into disuse. He refined the system of epicycles until they agreed so well with observations that nothing supplanted them for 1,400 years.

Nicolas Copernicus, in 1473, noticing that in the Ptolemaic system there are many *identical* epicycles, realized that they can all be eliminated if the Earth is made to go round the Sun. They are merely traces, superimposed on all the other bodies, of the Earth's motion. In other words, a heliocentric theory is *simpler* and thus should be preferred. As is well known, this suggestion did not meet with approval from the religious authorities. Copernicus wisely withheld publication until after his death. But later a more foolhardy Galileo was hauled up before the Inquisitor in Florence:

Propositions to be Censured. Censure made in the Holy Office of the City, Wednesday, February 24, 1616, in presence of the undersigned Theological Fathers. *First*: The sun is the centre of the world, and altogether immovable as to local movement. *Censure*: All have said that the said proposition is foolish and absurd in philosophy, and formally heretical, in as much as it expressly contradicts the opinions of Holy Scripture in many places and according to the common explanation and sense of the Holy Fathers and learned theologians. *Second*: The earth is not the centre of the world and is not immovable, but moves as a whole, also with a diurnal motion. *Censure*: All have said that this proposition must receive condemnation in philosophy; and with respect to theological truth is at least erroneous in faith.

Tycho Brahe (1546–1601) didn't believe the heliocentric theory either, but his interests in astronomy quickly became almost entirely experimental. He made extensive and remarkably accurate observations over a lengthy period, and eventually produced his own version of Copernicus's theory. On his deathbed he begged Johannes Kepler to sort out cosmology. Kepler felt that the Ptolemaic theory, even as revised by Copernicus, was dissatisfyingly arbitrary and complicated. He sought patterns in planetary motion. Not all of his patterns would meet with modern approval. For example in the *Mysterium Cosmographicum* he explains the spacing of the planets in terms of regular solids fitted between their spheres, thus:

Mercury
          Octahedron
    Venus
          Icosahedron
    Earth
          Dodecahedron
    Mars
          Tetrahedron
    Jupiter
          Cube
    Saturn

One advantage of this delightful theory is that it explains why there are only five planets (apart from the Earth which, of course, is different): there are only five regular solids to go round. At the time it seemed at least as sensible as anything else Kepler suggested (and it reminds me uncomfortably of modern theories of elementary particles in terms of representations of groups). Almost buried in Kepler's work are three fundamental laws, which have withstood the ravages of time much better.

He selected Mars to work on, a lucky choice as we shall soon see; and began by assuming that its orbit is an eccentric circle. Eventually, on this hypothesis of a circular orbit, he got his *Second Law*: a planet sweeps out equal areas in equal times. Then he decided that a circle wouldn't do, but some sort of oval might. Eventually the observations forced him, somewhat unwillingly, to his *First Law*: a planet orbits the Sun in an ellipse, with the Sun at one focus. It's here that the luck comes in: Mars has an especially non-circular orbit, so the elliptical shape shows up well in observations. Kepler then found that the second law still works for elliptical orbits, a discovery that impressed him no end. What about the distances of the planets from the Sun? As well as the regular solids, Kepler tried musical analogies, but with no better success. Finally he hit on the *Third Law*: the cube of the distance is proportional to the square of the orbital period. Kepler was over the Moon:

It is not eighteen months since I got the first glimpse of light, three months since the dawn, very few days since the unveiled sun, most admirable to gaze, burst upon me. Nothing holds me; I will indulge my sacred fury; I will triumph over mankind by the honest confession that I have stolen the golden vases of the Egyptians to build up a tabernacle for my God far away from the confines of Egypt. If you forgive me, I rejoice; if you are angry, I can bear it; the die is cast, the book is written, to be read either now or by posterity. I care not which; it may well wait a century for a reader, as God has waited six thousand years for an observer.

Although his three laws are almost buried in a mass of metaphysical speculation, Kepler had struck gold, and he knew it.

Equations for everything

Newton explained all this with his theory of universal gravitation. Every particle in the universe attracts every other particle by a force proportional to their masses and inversely proportional to the square of the distance between them. The 'inverse square' prescription is what gives Newton's theory its precision; but the universality—that it holds for *every* particle—is what gives it its power. It can be applied to any configuration of particles and bodies whatsoever. At least, in principle: it's easier to state an equation than to solve it. For two particles the equation can be solved, yielding elliptical orbits that obey the three laws discovered empirically by Kepler. Conversely, Kepler's laws show that if a force is the cause of planetary motion, then it must obey an inverse square law. Very fine effects in the motion of the Moon can also be explained by Newton's theories. Another, later triumph was the explanation of a grand cosmic dance in which Jupiter and Saturn alternately fall behind and ahead of their expected places. Even so, Newton's ideas at first met with a lot of opposition because the notion of a force acting at a distance was felt to be discredited.

More generally, people began to make use of *partial differential equations*, involving rates of change with respect to more than one unknown quantity. In 1746 d'Alembert applied them successfully to the vibrations of a violin-string. Daniel Bernoulli and Euler extended the ideas to water waves, sound waves, fluid flow, and the elastic deformation of solids. The *wave equation* became a standard item in the mathematical repertoire. Legendre and Laplace developed Newtonian gravitation into Potential Theory, studying such problems as the gravitational attraction of ellipsoids, it being known by then that the Earth was not a perfect sphere, but flattened at the poles. New astronomical discoveries, such as the asteroids, Uranus, and Neptune, fitted Newton's laws perfectly—and were predicted using them. Joseph Fourier's *heat equation* paved the way for thermodynamics and Fourier analysis. James Clerk Maxwell summed up all of electricity and magnetism in his *Maxwell equations*. Soon almost all of physics became formulated in terms of partial differential equations— to such an extent that even today, many scientists find it hard to conceive of any other way to model reality.

The Oscar award

As a result of all this activity, people began to think of Newton's description of planetary motion as some kind of Ultimate Truth. The Solar System formed a kind of infinitely accurate clock, set going when the planets were hurled by the

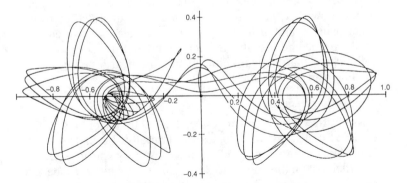

Fig. 22.   Orbit in a three-body system.

Hand of Creation. It was a clock that could never run down. But could it run wild? For instance, could the other planets resonate with the Earth like a child's legs pumping a swing, to make it collide with Mars, or run off into the cold and empty interstellar wastes? For imaginary solar systems simpler than ours, this kind of thing can happen. Could it happen to *us*? In 1887 King Oscar II of Sweden offered a prize of 2,500 crowns for an answer.

The state of the art at the time was far from adequate. It gave an exact solution in closed form for *two* bodies: Keplerian ellipses. A solar system having one sun and one planet could not run wild; all it could do was repeat the same motions over and over again forever. The general problem of *three* bodies seemed totally intractable. As far as we know today, no closed form solutions exist; at any rate, the general behaviour is enormously complicated. A few special solutions are known; for example Joseph-Louis Lagrange found that all three bodies can move in synchrony in elliptical orbits. But this type of motion is an accident of the mathematics, unlikely to occur in practice. Power series methods à la Newton help a bit, but are best on problems where most of the bodies have a mass small enough to be neglected: systems like the Sun, Jupiter, and an asteroid. Lagrange and Laplace between them did manage to show that the total departure from circularity of the orbits of the planets in the Solar System is constant; and that this total, even if concentrated on the Earth, will not switch it to an orbit that escapes the Sun altogether, like that of a comet. But this didn't show that the Earth might not slowly gain energy from other planets, and drift out by bigger and bigger almost-circles until we become lost in the silence and the dark. Proof was still lacking that we will neither leave the hearth nor fall into the fire, and for that the king offered his crowns.

The last universalist

Possibly the last person to understand fully *all* of the mathematics of his day was Poincaré. Naturally, he had a go at King Oscar's problem; and, characteristically, he invented a new field of mathematics to do it: *analysis situs*, now better known as topology. To see how it might apply, suppose that we could show that at some specific time from now, all of the planets would be in exactly the same relative positions as they now are. Then we would conclude from Newton's Law that the motion will thereafter repeat indefinitely. Poincaré saw that the essence of this idea is geometrical. According to Newton's dynamics, the motion of a system can be described if we know, at any instant, the positions and velocities of every particle in it. One way to imagine this is to invent an enormous multidimensional space, with six dimensions for each particle: three of position, three of velocity. The motion of a system of N bodies (say $N = 10$ for the Solar System, neglecting moons, or about 40 if moons are included, or billions if every particle in Saturn's rings is taken into account) corresponds to the motion of a single point in a $6N$-dimensional *phase space*. If the track in phase space joins up with itself to form a *closed* loop, then King Oscar's question is answered. No planet can ever wander off; the representative point, and hence the entire system, is *periodic*, going through the identical motions forever.

We thus see the importance of very simple-looking questions, such as 'does the path traced in phase space close up?' Poincaré's contribution was to find ways to tackle such questions *directly*, without first solving the equations or calculating the numbers involved. He did not settle the question of the stability of the solar system: that had to wait until the 1960s. But he made such a dent in it that in 1889 he was awarded his coveted Oscar, and he thoroughly deserved it. He had invented a whole new kind of mathematics. He showed how to reduce various physical problems to topology, and in some cases solved the resulting topological problems too. For instance, he proved that in the motion of three bodies there are always infinitely many distinct periodic motions, provided a certain topological theorem holds. In 1913, the year after Poincaré died, George Birkhoff proved the theorem.

The Great Escape

If the motion of the Moon can be predicted accurately, it is possible to navigate by observing its motion against the background of stars. So practical is *this* problem that a substantial sum of money was on offer for its solution. For this and other less mercenary reasons a lot of effort was devoted to the three-body problem: the motion of a system consisting of three point masses (such as Sun, Earth, Moon) moving under Newtonian gravitation. It's easy enough to write down the appropriate equations of motion; but immensely harder to solve them. A simple closed form solution like Keplerian ellipses doesn't exist. One practical method is to approximate the true motion by series. This can be done effectively, but the formulae get very complicated. Charles Delaunay devoted an entire *book* to one formula for the Moon's motion. Another method is to solve the equations numerically on a computer, but that's best for special problems, like landing on the Moon, and less helpful for general understanding.

As an aside: it has been said that one can gauge the progress of science by the value of n for which the n-body problem *cannot* be solved. In Newtonian mechanics the 3-body problem appears to be insoluble. In Relativity, it is the 2-body problem that causes trouble. Quantum Theory gets hung up on the 1-body problem (a particle); and Relativistic Quantum Field Theory runs into trouble with the 0-body problem (the vacuum)!

Messy calculations get answers, but often provide little insight. The modern approach to the 3-body problem is topological: its aim is to derive the main qualitative features. It is now known that these can be extremely complicated. For example a light body orbiting a pair of heavy ones can perform virtually any sequence of turns round first one, then the other.

Just how complicated the problems are is demonstrated by recent work on *singularities* in the n-body problem: events beyond which the motion cannot be continued within the mathematical model provided by Newton's laws. The simplest singularities are collisions, where several bodies meet. In 1897 Paul Painlevé proved that for three bodies, the only possible singularity is a collision. A two-body collision can be 'regularized' by making the bodies bounce elastically off each other: in this way a solution valid for times after the collision may be obtained. A simultaneous collision of three bodies cannot be regularized in this way: as far as the Laws of Motion go, such a collision might as well be the end of the universe, for there is no way to predict how the system will behave afterwards. Painlevé suggested that with more than three bodies there may exist much nastier types of singularity—for example bodies escaping to infinity in finite time, or oscillating ever more violently.

In 1984 Joseph Gerver (following earlier work by H. J. Sperling, Donald Saari, John Mather, and R. McGehee) gave a non-rigorous, but convincing, argument for the existence of configurations of five point masses which, under their mutual Newtonian gravitational attraction, fling themselves out to infinity in a finite time. The proposed mechanism is ingenious. Imagine three massive stars. (The astronomical terminology is just to make it easier to remember the orders of magnitude of the masses: I repeat that the bodies are *points*.) Arrange all three stars in a triangular constellation with an obtuse angle at the heaviest one, and start them out in such a way that the triangle expands at a uniform rate while maintaining its shape. Add a tiny asteroid which orbits round the outside of all three, approaching them very closely. As the asteroid passes each star it undergoes the same 'slingshot' effect that was used by the Voyager spacecraft as it journeyed through the Solar System. As a result, it gains a small amount of kinetic energy from one star, which it can then share with the other two to increase their speeds (and its own). If only there were a way to increase the energy of the first star as well, it would be possible to make the triangle expand at an ever-increasing rate, disappearing to infinity in a *finite* time. Of course, the law of conservation of energy forbids this. But there's a way out. Add a fifth body, a planet orbiting the heaviest star. Now make the asteroid gain energy at the expense of the planet, and transfer some of it to its star via a zig-zag slingshot. Otherwise, proceed as before. Now, as time increases, the constellation expands ever faster but maintains its shape; the asteroid whizzes round ever more rapidly in its triangular orbit; and the orbit of the planet descends ever closer to its star. As a result, *all five bodies* escape to infinity in a finite time.

It's a nice idea, but it involves a very delicate balancing act. You've got to arrange for an infinite sequence of slingshots to take place without disturbing the general scenario. Gerver's original arguments on this point lacked rigour—as he himself pointed out—and the detailed calculations became so messy that the proof couldn't be pushed through. However, stimu-

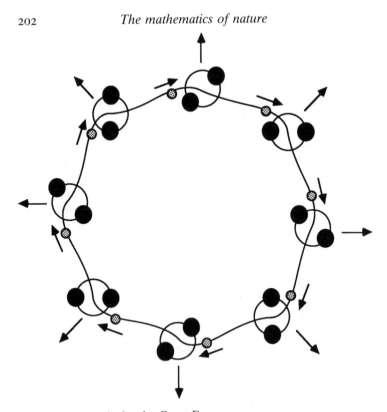

Fig. 23.   Scenario for the Great Escape.

lated by Gerver's new ideas, Z. Xia proved in 1988 that the 5-body problem really does have a solution in which all five bodies escape to infinity in finite time. His scenario is different from Gerver's, and requires three dimensions of space rather than the two that suffice to set up Gerver's system.

By 1989 Gerver had come up with yet another approach, not requring a third dimension. He used a symmetry argument suggested by Scott Brown, to prove that the 3*n*-body problem permits an escape to infinity for large enough *n*. Symmetry makes the calculations tractable—though still far from easy. The configuration required is a descendant of the triangle of stars. Take *n* pairs of binary stars, of equal mass. Each pair rotates around its mutual centre of gravity in a nearly circular orbit, and the centres of mass of the pairs lie at the vertices of a regular *n*-gon. A further *n* planets, whose masses are equal, but much smaller than the mass of the stars, follow orbits that approximate the edges of the polygon. Whenever a planet approaches a binary star, it gains kinetic energy via the sling-shot effect, while the binary star loses kinetic energy and moves to a tighter orbit. The planet also transfers momentum to the binary pair, causing it to move outwards, away from the centre of the polygon. Because of symmetry, all pairs are affected in exactly the same manner at exactly the same time. At each slingshot, the polygon grows, the planets speed up, and the binary stars close up into tighter orbits.

By suitably adjusting the number of bodies, their masses, and their initial positions and velocities, infinitely many sling-shot events can be made to occur during a finite time. The polygon expands to infinity, while the encircling binary stars oscillate more and more wildly as their orbits shrink to zero radius.

Symmetry effectively reduces the problem from $3n$ bodies to three. Once we have determined the positions and velocities of one binary star and one planet, the *n*-fold rotational symmetry determines those of the remaining $3n - 3$ bodies. Thus we can keep track of just three bodies, the rest being taken care of by symmetry. In other words, the problem reduces to a different problem about three disconnected 'bodies' (each an *n*-gon of point masses) moving under rather complicated forces. This problem is tractable, at least for large enough *n*, although the calculations remain far from easy.

The Great Escape is purely a mathematical phenomenon. In the real world particles aren't points, so the binary stars would eventually collide. Moreover, real velocities are limited by the speed of light, so nothing can *ever* escape to infinity in finite time.

Is *that* why the deity made the universe relativistic?

The stratosphere of human thought

After its impressive beginning, with Birkhoff's proof of Poincaré's conjecture on the 3-body problem, topology rather faded from the view of the scientific public. This hap-pened because topological problems are *not* always easy to solve. Calculus got off to a good start because there are plenty of functions that you can easily differentiate or integrate, and plenty of differential equations that *can* be solved, even if the most interesting ones can't. Topology, picking up at the point where calculus got stuck, was deprived of easy conquests. For example you can give a nice topological proof that a pendulum oscillates periodically; but since this is already known (even though the usual proof is a fudge) nobody except topologists will be impressed by it. Because it was so difficult to solve the topological problems posed by physics, mathematicians were unable to act as short-order cooks for the physicists, with the motto 'bring us your problem in topological form and we'll tell you the answer'. So mathematicians had to start over, and develop a whole new chestful of tools for handling topological problems.

I don't intend to suggest that topology was isolated from other branches of *mathematics*: quite the contrary. In the 1930s and 1940s Lefschetz applied it to algebraic geometry, and Marston Morse to differential geometry (in work that has since become important in scientific applications). But topology wasn't widely seen as What Every Young Physicist Should Know. Indeed, by the 1950s, all of this activity had got pretty far removed from physics. It had to: that's the direction topology led. So topology turned inward upon itself, to solve its internal problems. Physicists promptly lost interest in it: to them it looked about as relevant as the sage Gautama sitting cross-legged under a tree meditating on the nature of suffering. To a non-specialist, the degree of introspection required to sort topology out appears extreme. Thus Alexander Solzhenitsyn, who had enough mathematical training to recognize this self-absorption, but lacked the sense of history to understand the reasons for it, wrote in *The First Circle*:

Nerzhin, his lips tightly drawn, was inattentive to the point of rudeness: he did not even bother to ask what exactly Verenyov had written about this arid branch of mathematics in which he himself had done a little work for one of his courses. Suddenly he felt sorry for Verenyov. Topology belonged to the stratosphere of human thought. It might conceivably turn out to be of some use in the twenty-fourth century, but for the time being . . .

> '*I care not for the sun and the stars,*
> *I see but man in torment.*'

But the modern prophet of Mother Russia has been proved wrong within his own lifetime. Almost everything interesting being done in modern dynamics involves a more or less heavy dose of topology, which has brought a semblance of order to what looked, twenty years ago, to be a hopeless muddle and a baffling enigma.

The world in a hairy ball

To see the way in which topological insight can illuminate a wide variety of apparently unrelated things, consider one of its minor triumphs, which says (in suitably refined language) that you can't comb a hairy ball to make it smooth everywhere. Here are some direct consequences:

(1) At any moment, there is at least one point on the Earth's surface where the horizontal windspeed is zero. (If not, comb along the wind.)

(2) A spherical magnetic bottle, as used in 1950s fusion reactor experiments, must leak. (It leaks where lines of magnetic force run perpendicular to the surface of the bottle, i.e. where the horizontal component is zero.)

(3) Dogs, bears, cheetahs, and mice must have a parting in their fur. (Self-explanatory.)

(4) Every polynomial equation has a complex root. (Technical proof based on the fact that the complex numbers, plus a point at infinity, form a sphere.)

(5) A chemical gradient on the surface of a cell must have a critical point (where there is no local 'downhill' direction).

This is a relatively simple case, and the 'applications' are not necessarily burning issues at the frontiers of science. But recent uses of more difficult and technical topological theorems most certainly are, notably in the current search for a way to unify the four fundamental forces of Nature (chapter 19). Think of it merely as an accessible example of the way topological reasoning can be all-embracing.

The advantages of navel-gazing

The subjects that originally provoked the creation of topology are no longer fundamental physics. The stability of a Newtonian Solar System is beside the point when the currently interesting one obeys Einstein's laws, not Newton's. And anyway, atomic theory says that the Sun will eventually blow up, making the long-term future of heliocentric planetary motion, either Newtonian or Einsteinian, moot. All of these far-reaching changes to science occurred over a period of some five decades, during which topology did little save contemplate its own navel. (Mind you, what subject is better fitted to do this?) Since the whole self-indulgent exercise's sole practical motivation was in Newtonian physics, it looks as though, while the doctor debated hypothetical points of treatment, the patient died. But paradoxically topology, now honed to a fine cutting edge, remains highly relevant to fundamental physics and other sciences. For example (2) is about nuclear engineering, (5) biochemistry. Not only did Poincaré not design topology with applications to these subjects in mind: the subjects themselves did not exist when topology was born.

How can this be? Was Poincaré a better prophet than Solzhenitsyn? Well, very likely, but that's beside the point. The current success of topology, some examples of which we examine in detail in later chapters, owes little to Poincaré's abilities at crystal-gazing, but an enormous amount to his mathematical imagination and good taste. Topology is a success precisely *because* it forgot the details of those original problems, and instead concentrated on their deeper structure. This deep structure occurs in any scheme of mathematical investigation that makes use of continuity. In consequence, topology touches on almost everything. Of course, the extent to which this fact helps depends on the question you want to answer.

Beware the galactic mouse!

We'll take a further look at topology's contribution in the next few chapters; but it would be unreasonable to close this one without giving the modern answer to King Oscar's problem. In 1963, using extensive topological arguments, Kolmogorov, Vladimir Arnol'd, and Jurgen Moser were able to respond to the question 'Is the Solar System stable?' with a resounding and definitive answer: 'Probably'. Their method (usually called KAM Theory) shows that *most* of the possible motions are built up from a superposition of periodic ones. The planets never *exactly* repeat their positions, but keep on almost doing so. However, if you pick an initial set-up at random, there is a small but non-zero chance of following a different type of motion, whereby the system *may* lose a planet (or worse)—though not terribly quickly. It all depends on exactly where, and at what velocity, the planets are placed by the Hand of Creation. The fascinating point is that there is no way to tell by observation which of these two types of behaviour will occur. Take any configuration that leads to almost periodic motion; then there are configurations as close as you like where planets wander off. Conversely, adjacent to any initial configuration from which a planet wanders off, there are others for which the motion is almost periodic. The two types of behaviour are mixed up together like spaghetti and bolognese sauce. If a mouse on a planet a billion light-years away were to twitch one whisker, it could switch the Solar System from a stable configuration to an unstable one, or vice versa. It would be difficult to confirm this theory by direct observation, just as it is hard to confirm the universality of Newtonian theory. But KAM Theory does make predictions that can be tested observationally, for example about the spacing of asteroids round the Sun, and Saturn's rings. It also applies to Quantum Theory, for example the instabilities of particles in accelerators. Topology has revealed complexities that King Oscar never dreamed of; but it has also given us the power to tame them and exploit them.

16
Oh! Catastrophe!

> Many phenomena of common experience, in
> themselves trivial—for example, the cracks in
> an old wall, the shape of a cloud, the path of a
> falling leaf, or the froth on a pint of beer—are
> very difficult to formalize, but is it not possible
> that a mathematical theory launched for such
> homely phenomena might, in the end, be more
> profitable for science?
>
> René Thom

It is all too easy to assume that gradually changing causes must
produce gradually changing effects. Why did the dinosaurs die
out so suddenly? One current theory invokes a massive meteor
impact on the Earth. Others propose dramatic changes in
world climate, or epidemics of disease, or loss of food sources.
The tacit assumption is that a dramatic effect should have a
dramatic cause. It would be surprising if, in a world where
nothing much seemed to be changing, all the dinosaurs suddenly took it into their heads to become extinct.

And yet . . . heat a beaker of water, gradually, steadily. It
warms up equally gradually and steadily—until the temperature hits 100 °C. Then, with no change in the rate of heating, it
suddenly begins to boil, and turns to steam. Why? What
dramatic cause produced this literally explosive effect? What
was special about the steam-producing heat? What special
feature of straw $N + 1$ causes the camel's back to break, when
the previous N have no effect? Why do hydrogen and oxygen
react gently at one pressure, but explode at another, possibly
lower, pressure? Why do balloons burst, ships capsize, elastic
bands snap, cornered rats attack, and harassed fathers lose
their temper? Gradually changing causes *can* produce sudden
effects. But how?

A mixed bag

The traditional calculus-based mathematical models employed by science are at their best when describing smooth, continuous changes. The newer methods of combinatorics and computer science—the vogue term is 'finite mathematics'—are best suited to phenomena that are consistently discrete. But many of the most interesting natural events display a mixture of the discrete and the continuous. The smooth wave that rolls up a gentle beach and breaks in a shower of spray; the swaying branch of a tree that sways a fraction too far and snaps; the smooth growth of an embryo, punctuated by dramatic changes in form as rudimentary organs begin to appear. One way to study such events is to use either a discrete or a continuous model, as appropriate. This is a rather *ad hoc* chewing-gum-and-string approach, and as such is aesthetically unsatisfying. Nature often seems to integrate the two types of change as if they are part of a single process. So it is an attractive idea to develop a comparable mathematical theory that can deal in a natural, organized, and unified way with systems that usually change smoothly, but sometimes jump. Such was the aim of René Thom when, in 1972, he published *Structural Stability and Morphogenesis*, the manifesto for a new type of mathematics now known as catastrophe theory (or, in some of its aspects, singularity theory). It is called this, not because it deals in disasters, but because Thom used 'catastrophe' in its original, Greek sense of an unexpected, dramatic change.

Thom is not the only person to have thought about the problem, and a number of other schools of thought provide further sources of ideas. They include classical bifurcation theory, which studies the 'branching' of solutions to differential equations; topological dynamics; the theory of co-operative phenomena, otherwise known as synergetics; and non-equilibrium thermodynamics. All have their part to play, and I suspect that the various approaches will eventually be unified as the viewpoints acquire wider currency; but for simplicity I shall confine my account to the mathematics provoked by Thom's ideas.

The origins of catastrophe theory, in retrospect, can be traced back a long way—for example to Archimedes's *On*

Floating Bodies, the work of Descartes and Huygens on the rainbow, and Leonhard Euler's theory of the buckling beam. The subject evolved gradually, until in the late 1960s it underwent a sudden and dramatic explosion (mimicking the very pattern of behaviour it attempts to describe). Thom's work caught people's imagination, because it brought together the previous strands into a coherent whole. He saw, and talked about, the Big Picture. And he orchestrated its development. He was not alone; in particular Arnol'd developed very similar ideas, and Bernard Malgrange and Mather supplied rigorous proofs of several basic theorems, adding their own generalizations. Thom was a topologist, and his main stated objective was developmental biology, but his starting-point was a vague analogy with a mathematically more precise subject: optics. His book is condensed, idiosyncratic, uncompromisingly blunt, and of a philosophical cast which many scientists find annoying. But buried in it are nuggets of mathematical gold which, duly extracted and refined, are proving valuable in a range of sciences from engineering to psychology.

The landscape of change

The biologist C. H. Waddington was interested in the way organisms develop definite forms as they grow—*morphogenesis*. A developing embryo appears to go through periods of gradual growth, punctuated by sudden events such as the budding of a limb or the differentiation of nerve cells from muscle. Waddington devised a 'hydraulic analogy': water flowing through a smoothly undulating landscape. The water, seeking the lowest point, collects in pools. Each pool fills gradually, until the water reaches its rim, but then it sudddenly overflows and new pools start to form elsewhere. Despite the smoothness of the landscape and the steady supply of water, occasional discontinuities occur. An embryo might flow in a similar way through a multidimensional biochemical 'landscape', with overflows acting as triggers for morphological change.

One way to reformulate this mathematically is to think of the landscape as the graph of a smooth function, which is slowly changing with time, and to ask how the local minima (valley-bottoms, where the water can collect) can vary. If you want to know very detailed features, such as the distances between pools, the shapes of the pools, or the amount of water in each, then the answer is 'rather a lot', and you don't get very far. But if you're interested only in the qualitative, topological features—for example, how many pools a given one splits into when it overflows—then, according to Thom, very few things can happen.

The Magnificent Seven

As one example of the kind of ideas he had in mind, Thom stated that in any system that seeks to minimize a function (height, in the hydraulic model, but it might be energy, complexity, cost, likelihood, or whatever you please), then only seven different local forms typically occur when the changes in the landscape are controlled by four or fewer variables. The word 'typically' refers to a mathematical feature of the classification: it excludes certain 'infinitely rare' exceptions. In much the same sense a 'typical' coin toss is either heads or tails, but not on edge; a 'typical' point on a dartboard is not the exact centre, a 'typical' member of the public is not the President of the United States. The atypical *can* happen, but if it does, some special explanation is warranted. (And a modified mathematical setting may then be in order.) The typical forms are known as the *fold, cusp, swallowtail, butterfly, elliptic umbilic, hyperbolic umbilic,* and *parabolic umbilic* catastrophes. Each has its own distinctive and beautiful geometry. There is a lot more to catastrophe theory than just the classification of the seven elementary catastrophes, but they will serve to illustrate the unifying power of the ideas.

The simplest elementary catastrophe is the fold. Here the sudden change is caused by a single control parameter exceeding some specific threshold value. Imagine a small dent on the slope of a large valley, containing a pool of water. Now gradually flatten the dent out. At a specific moment, the water will overflow and trickle down the side of the valley to the bottom.

The cusp catastrophe requires two control parameters. You can think of it as describing the way two nearby pools can merge or disappear as their depths vary. If one shrinks while the other deepens, the deeper one will eventually gobble up the shallow one. Since we can choose to make either one

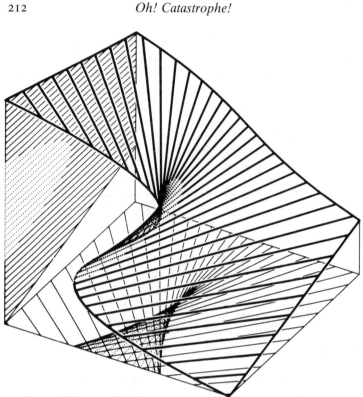

Fig. 24.   Cusp catastrophe surface, ruled by straight lines.

deeper, each can gobble up the other. On the other hand, after the gobbling is finished, all we have is a single pool, and we can't tell which one it was originally. We can even start with one pool, make it extrude a small one, then deepen the small one and shrink the deep one until the new pool gobbles up the original; after which we change the shape and position of the new pool until it looks exactly the same as the one we started from. Clearly the question 'Which pool is which?' requires some careful thought under these circumstances. In particular, you can*not* decide it merely on the basis 'Does the change from one to the other occur suddenly or smoothly?'

The boiling of water is a cusp catastrophe. One pool is

'liquid', the other 'gas'. If 'gas' gobbles up 'liquid' then the water boils—suddenly. If 'liquid' gobbles up 'gas' then the steam condenses. But—as every budding physicist knows—a suitable sequence of variations of pressure and temperature can cause a liquid to change *continuously* into a gas, without ever really boiling. So one pool can assume the identity of the other, in a perfectly continuous way. Many models of 'conflict' between two states similarly boil down to a cusp.

The elementary catastrophes (not to mention their less elementary brethren) arise throughout the sciences. The fold catastrophe, for example, occurs in the optics of rainbows and the formation of sonic booms. The cusp may be found in models of capsizing ships, boiling fluids, the onset of magnetism, optical focusing, buckling, chemical explosions, fluid flow, population growth, and hyperthyroidism. The swallowtail governs the motion of an elliptical roller on an inclined plane. The butterfly appears (appropriately) in the stability of aircraft. The hyperbolic umbilic describes the buckling of a stiffened plate. The elliptic and parabolic umbilics appear in the optics of liquid droplets. All of these applications involve standard, accepted scientific models and equations, but there are also a number of more speculative suggestions which require further work before they may be considered established. To give a flavour of the two types, I'll describe one of each. To reflect Thom's motivation, I'll take one from biology, and one from optics.

Sunfish invaders

In Lake Opinicon, Ontario, lives a creature to delight the hearts of dedicated feminists—the pumpkinseed sunfish *Lepomis gibbosus*. For it is the male fish that builds and tends the nest, guards the eggs, and looks after the offspring. Of course, in typical male fashion, he is an aggressive little beast, but that's a result of his maternal instincts. Imagine father fish, in protective mood, watching an intruder approaching. As long as the intruder stays far enough away, it will remain unmolested. But if it gets closer to the nest than some critical threshold distance, father will chase it away in a display of aggression. The most dangerous intruder, of course, is another fish of the same species.

Oh! Catastrophe!

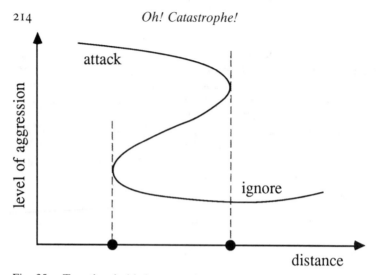

Fig. 25. Two thresholds in aggressive response to an invader.

A catastrophe-theoretic model of this behaviour was proposed by Christopher Zeeman in the early 1970s. It involves a
cusp catastrophe, with the two conflicting states being the
protective mood and the aggressive mood of the fish. It leads
to the prediction that there will be two threshold distances, not
just one. If an invader crosses the inner perimeter, then father
will attack. But only after it has crossed the outer perimeter
does he call off the attack. There is a 'buffer zone' between
the two perimeters. A fish within the buffer zone is safe; but if
it should cross the inner perimeter it will be chased, not just
into the buffer zone, but outside it. It's really a very pretty
mechanism. The biologists P. W. Colgan, W. Nowell, and N.
W. Stokes tested this theory, wading in rubber boots at the
edge of Lake Opinicon with a small wooden dummy invader.
They found that during the nest-building season, the average
inner perimeter is 13 cm., the outer 18 cm. This is a significant
difference. However, during the mating season the difference
shrinks to zero and the mood of the fish ceases to be either of
the protective or aggressive extremes. As you will appreciate,
this is important, since an approaching female should not,
under these circumstances, be chased away.

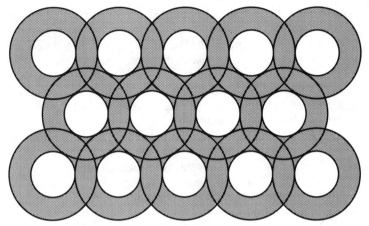

Fig. 26.   Buffer zones created by having two thresholds.

A final bonus is a result not envisaged in the original theory, but seen to follow from it. The existence of the buffer zones between close-packed nests leaves a kind of urban swimway system through which all fish can travel in search of food. If there had been a single perimeter, the nests would have been packed together without gaps, like soap-bubbles in a sink. Any fish attempting to leave its nest would be under permanent attack, with no time to seek food; and the species would die out at once. This raises a speculative possibility, that evolutionary pressures may have produced the two-perimeter effect.

Twinkling starlight

Zeeman's model of aggression does not derive its catastrophe mathematically from pre-existing principles: instead it takes the catastrophe as a working hypothesis and derives conclusions, which are then tested experimentally. But it is also possible to *use* catastrophe theory as a mathematical tool, deriving new consequences from known principles. Optics is a case in point. When light rays pass through a system of lenses and mirrors, they are often focused to form regions of unusual brightness, known as *caustics* because heat, as well as light, is concentrated there. Catastrophe theory enters because there is a fundamental principle, going back to Fermat and placed on a firm basis by Hamilton, that light selects which path to follow by attempting to minimize the travel time. The mathematical landscape that describes the physical system is a fictitious space whose points represent possible paths, and whose 'height function' is travel-time. The valley-bottoms are then 'points' of lowest 'height', that is, paths of shortest time. The upshot of the theory is a list of all possible forms for caustics, under suitably typical conditions.

The rainbow is an example of a fold catastrophe. Light travelling through a raindrop turns through different angles, depending on whereabouts it hits the drop. At a certain angle the deviation of the emergent rays 'turns back' in a mathematical fold, and light accumulates along the fold, to produce a bright luminous cone in the sky. Each colour of light has its own cone, and our eyes assemble sections of these cones to form concentric circular rings of colour. A rainbow seen from an airship is a complete circle: one appeared in the film *The Hindenburg*. Jumbo jets fly too high to see rainbows.

To observe an optical cusp catastrophe, look into a full cup of coffee on a sunny day, or with a bright light off to one side. You will see a bright curve with a sharp point, or cusp, on the surface of the coffee. This is where rays of light, reflected from the sides of the cup, bunch together.

There is at least one optical effect that brings together *all possible* catastrophes. It is the twinkling of starlight. This is caused by random fluctuations in the Earth's atmosphere, which acts like an irregular, moving lens. The starlight is focused and unfocused at random, producing a rippling pattern of caustics passing across the observer's eye. Each catastrophe concentrates light in its own distinctive way, so different types of caustic lead to different degrees of brightness. The rarest catastrophes (of large co-dimension) are also the brightest, and the twinkling balances these two effects.

The physicist Michael Berry has used catastrophe theory to calculate what is called the *short-wave asymptotics* of twinkling. He assumes that the random fluctuations of the atmosphere obey the statisticians' favourite rule, known as *Gaussian statistics*. Because of the complex competition between catastrophes in the caustics, the intensity of the observed starlight, which also varies randomly, does not change in such a nice way. Its statistics are different from those of the atmosphere, and are very far from Gaussian. The brightness also depends strongly on the wavelength of the light. The basic question is: how do the statistics of twinkling vary as the wavelength becomes small? The most important statistics are called *moments*. The nth moment is the average value of the nth power of the light intensity. In principle, if you know all the moments, you know everything there is to know about the statistics. In practice the first few moments provided important coarse-grained information. The first moment is just the average; the second tells how much the fluctations spread out around this average; the third measures how lopsided the spread is, and so on. Berry proves that each moment varies as a fixed power of the wavelength, called a *critical exponent*, and uses the classification of catastrophes (as extended by Arnol'd) to compute them. Each is a rational number. In three dimensions, the list from moments 2 to 13 is:

0, 1/3, 1, 5/3, 5/2, 7/2, 9/2, 11/2, 13/2, 38/5, 87/10, 157/16.

Even though catastrophe theory is topological, hence 'qualitative', these are *numerical* predictions. The first few have been tested experimentally and the results agree with the theory. Testing the higher moments is more difficult because the equipment sensing the light must respond very rapidly to large changes. This application sheds an interesting light on the assumption that, because topology is qualitative, it cannot be used to make quantitative predictions. It also shows that the classification theorem for catastrophes is not merely a taxonomic device.

Variations on the theme

Although the elementary catastrophes can be completely classified, this is not the end of the story, because they occupy a very special niche in mathematics. Their importance for mathematics is not so much as an end in themselves, but as a means: their theory provides a paradigm upon which future work can be based. It tells us what to do to get started. Many more general 'catastrophes' can be considered, and the theory in any case has more than classification to offer. Current research in catastrophe theory centres around a generalization conceived in 1978 by Martin Golubitsky and David Schaeffer. They realized that in many applications there is a 'distinguished parameter', and the important question is: how does everything vary as that parameter changes? For example, in a chemical reaction we want to know how the reaction rate varies with temperature, so the distinguished parameter is temperature; in engineering we want to know how a strut behaves under load, so the distinguished parameter is load. Incorporating this new ingredient requires some technical changes to the mathematics, and the detailed results are, of course, different; but the general pattern of the mathematics is exactly the same and so are the methods of proof. It's a bit like driving a car: if you've learned on a Ford then you'll have little trouble driving a Volkswagen, even though many of the controls are in different places. Golubitsky and Schaeffer's theory has led to a large number of new applications to chemical reactions, fluid flow, geophysics, and elasticity. Further, generalizations are emerging at a steady rate, along with new applications.

Many natural and experimental systems possess symmetry: for example, planets are approximately spherical, molecules of methane are tehrahedral, and pipes are cylindrical. The Golubitsky–Schaeffer theory has been extended to take symmetry into account. As a result, it provides a general framework in which to study a phenomenon known as *symmetry-breaking*. A general principle enunciated by the physicist Pierre Curie states that the symmetries of causes are inherited by their effects. If interpreted in the most obvious manner, however, this statement is false. For example, imagine a spherical shell compressed by a uniform force. The shell is spherically symmetric, and so are the forces (cause); but if the forces exceed a critical value, the shell buckles into a pattern that is *not* spherically symmetric. In fact, the first buckle leads to a shape with rotational symmetry about some axis.

These ideas have been applied to a great variety of problems. For example, if a thin layer of fluid is heated from below, a pattern of convection cells is formed. It may consist of parallel rolls or hexagons arranged like a honeycomb. Similar effects occur if a layer of fluid confined to the region

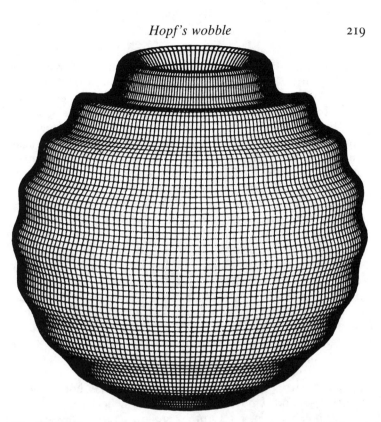

Fig. 27. When a sphere first buckles under uniform pressure, it loses its spherical symmetry but retains circular symmetry about some axis. (J. M. T. Thompson and H. B. Stewart, *Nonlinear Dynamics and Chaos*, John Wiley and Sons Ltd.)

between two concentric spheres is heated; now the cells form polyhedral patterns. This system models convection in the Earth's mantle. Other applications include cellular flames and elastic buckling.

Hopf's wobble

Another trend is to make the theory more dynamic, so that it can deal with oscillations as well as steady equilibria. The basic example here is known (after its discoverer Eberhard Hopf) as *Hopf bifurcation*, in which a steady state becomes unstable and turns into a periodic oscillation. Ralph Abraham calls it the 'wobble catastrophe'. It turns out that the mathematics of Hopf bifurcation can be reformulated as catastrophe theory plus circular symmetry. Using this, a new theory of degenerate Hopf bifurcation, more general than Hopf's original theory, has been developed by Golubitsky and William Langford. Applications include work by Isabel Labouriau on the transmission of nerve impulses. Hopf bifurcation with symmetry, and very recently degenerate Hopf bifurcation with symmetry, can now be handled by these methods.

Applications include chemical oscillations, vibrating pipes, and even the patterns in which animals move their feet when they walk, trot, run, gallop, and so on. One laboratory system that has been extensively analysed by these methods is a famous experiment in fluid dynamics due to G. I. Taylor in 1923. The apparatus consists of two circular cylinders, one inside the other. The space between is filled with fluid, and the cylinders are rotated. What happens? One might imagine that the fluid just rotates evenly as well, but this is only true at low speeds. As the speeds increase, Taylor vortex cells form in horizontal layers. Then the cells develop a wave (wavy vortices) which itself rotates rigidly. The wave then starts to go up and down (modulated wavy vortices) like horses on a merry-go-round. Under other conditions helical vortices, in a barber-pole pattern, can form. There are wavy helices, twisted vortices, braided vortices, modulated wavy helices. It now appears that virtually all of these effects can be explained in terms of broken symmetries; and it is the circular symmetry of the apparatus, rather than the equations of fluid flow, that is responsible for them. A mathematical theory that can shed new light on a problem studied by hundreds of scientists for more than sixty years can't be all bad.

The origins of catastrophe theory are part mathematical, part physical, and mostly biological. It developed from fairly vague analogies, many of which do not withstand severe scrutiny. But motivation and results are two different things. It has led to a fundamental and extremely important set of mathematical ideas; a number of speculative applications to the 'soft' sciences which have yet to find favour with their practitioners; a series of unexpected generalizations now having a major impact on the main competitor, bifurcation theory; and a wide range of successful applications to subjects not originally envisaged by anybody. It resembles the phases of matter: in the physical sciences it is solid, in biology somewhat fluid, in sociology possibly gaseous. Whether it is developing the way its creators expected is moot. But it is certainly developing into a powerful set of ideas. And, as an example of how mathematics conceived for one purpose can acquire applications to many others, it shows how curious, indirect, and sometimes downright perverse the influence between different branches of science can be.

17

The patterns of chaos

Chaos of thought and passion, all confused;
Still by himself abused, or disabused;
Created half to rise, and half to fall;
Great Lord of all things, yet a prey to all;
Sole judge of truth, in endless error hurled;
The glory, jest, and riddle of the world!

Alexander Pope

'Rough winds', said Shakespeare, wondering aloud whether to compare thee to a Summer's day, 'do shake the darling buds of May.' We get the rough winds most years; but May 1985, for example, will be remembered mostly for being *cold*. In 1984 our ancient apple-tree produced over 400 lb. of apples, ending up as 18 gallons of apple wine, dozens of pies, pounds of purée, and finally compost. A year later an Easter snow-storm damaged either buds, or bugs, or both, and we got a miserable 20 lb. I'd hate to guess about this year's crop: British weather is notoriously unpredictable. But what do we *mean* by that?

The Age of Reason, which was Newton's legacy, viewed the System of the World—that is, *everything*—as some gigantic and intricate clockwork machine. If you knew enough about the way the machine fitted together, you could in principle say what was going to happen from now till the crack of doom. In practice nobody could predict what would happen to a mere three bodies under the action of gravity, but grand philosophical positions are not renowned for taking such material facts into account. The mathematical foundation of this view was that the motion of everything in the universe is described by some differential equation; and if you know the 'initial conditions' then all subsequent motion is fixed uniquely. So why can't you predict the result of rolling a pair of dice? Because you don't know enough about the initial conditions. If you rattle the dice in a cup, you have no idea what's happening to them. But once they hit the table, it would be possible to take a series of rapid snapshots, work out the positions and velocities, stick the whole thing on a supercomputer, and—if the computer were fast enough—work out which faces would end up on top, *before* the dice worked it out for themselves.

Determinism

Laplace stated the viewpoint clearly.

An intellect which at any given moment knew all the forces that animate Nature and the mutual positions of the beings that comprise it, if this intellect were vast enough to submit its data to analysis, could condense into a single formula the movement of the greatest bodies of the universe and that of the lightest atom: for such an intellect nothing could be uncertain; and the future just like the past would be present before its eyes.

This is the classical view of *determinism*. In principle, the entire history of the universe, past and future, is determined once you know exactly where every particle is going, and how fast, at one single instant of time. This leads to a delicate ethical and moral problem, that of free will. As Voltaire pointed out, 'it would be very singular that all nature, all the planets, should obey eternal laws, and that there should be a little animal, five feet high, who, in contempt of these laws, could act as he pleased, solely according to his caprice.' Does such a thing as free will exist? As a practical matter, even if it does not, its non-existence appears to constrain us *to act as if it did*. That is, when all the atoms in my body, following nature's immutable laws, cause me to mug an old lady, then those same immutable laws cause the magistrate to send me down for a five-year stretch.

A non-deterministic, or stochastic, system behaves quite differently. The same initial conditions can lead to different behaviour on different occasions. For example imagine a pendulum swinging in a random breeze. The equation describing the motion will have explicit random terms in it; and even if you set everything going from exactly the same starting-point, interference by the breeze will soon change the behaviour. Let me emphasize that in some sense the distinction is a modelling one. If we add deterministic equations describing the behaviour of the breeze, then the initial conditions for *that* also come into play, and the whole game becomes deterministic again. Indeed, it is impossible to prove any event is really random by repeating an experiment exactly: either time or place must differ, otherwise it's the *same* experiment. This again emphasizes the need to discuss the difference between determinacy and randomness within an agreed context, a specific model.

The extent to which practice fails to measure up to the deterministic ideal depends on how accurately the initial conditions can be observed (and, of course, on whether you're using the right differential equation). So, if you want to predict the weather mathematically, you need two ingredients:

(1) As accurate as possible a system of differential equations to describe the motion of the Earth's atmosphere.
(2) The best possible observations, using ground stations, ships, aircraft, meteorological ballons, and satellites.

After that, it's just a matter of computation, right? By now you won't be surprised to find it isn't quite that straightforward. For a start, if you make the differential equations *too* accurate, you end up modelling sound waves, not weather; moreover, these resonate with your numerical method, blow up, and swamp everything interesting. The solution is deliberately to coarsen the accuracy of the equations! But there's a much more serious problem. Topologists, following the lead of Poincaré in qualitative dynamics, asked whether a typical differential equation behaves predictably. To their surprise, Smale discovered that the answer is 'no'. Determinism is not the same as predictability. Indeed a perfectly deterministic equation can have solutions which, to all intents and purposes, appear random.

I'm not sure whether this solves the moral problem of free will, but it certainly puts it in a new light. You may recall how Douglas Adams, in *The Hitch-Hiker's Guide to the Galaxy*, tells of a race of hyperintelligent pan-dimensional beings who build a supercomputer called Deep Thought. When asked for the answer to 'Life, the Universe, and Everything', it announces that it can find it, but the program will take some time to run: about seven and a half million years. At the end of that time it announces the answer: 'forty-two'. Which leaves just one further problem: to work out what the question was. Laplace's hypothetical intellect looks about as practical a proposition to me; and its answers would probably be equally intelligible.

Familiar attractors

Topologists naturally think of a differential equation in a more geometric way than analysts do. They think of it as a flow. The state of the system is described by a point of a multi-dimensional phase space; and as time flows, so does the point. The path traced out in phase space by a moving point represents the evolution of the state of the system from a particular initial condition. The system of all such paths—the *phase portrait*—gives a qualitative picture of what happens to *every* possible initial condition. So the question is: what are the typical kinds of phase portrait?

The simplest systems are those that settle down to a steady equilibrium. Here all the paths converge on to a single point; that is, all histories end up in the same unchanging state. Topologists say that the point is an *attractor*. More generally an attractor is some region of phase space such that all nearby points eventually move close to it. The attractors describe the long-term evolution of the system.

The next simplest attractor is a *limit cycle*, where paths converge on to a closed loop. This corresponds to a periodic oscillation of the state of the system; periodic because the representative point keeps tracing out exactly the same sequence of events as it goes round and round the loop.

Finally there is *quasiperiodic* motion, in which a lot of periodic motions with different periods are superimposed. The Earth going round the Sun, the Moon round the Earth, a spacecraft orbiting the Moon, and an astronaut swinging a cat inside it—assuming there's room to swing a cat in a spacecraft. The cat's motion is compounded from four different periodic motions, with different periods. Over millions of years the cat will come very close to repeating its previous movements, but (unless all four periods are exact integer multiples of some fixed time interval) it will never repeat exactly. Topologically

The patterns of chaos

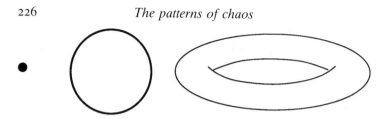

Fig. 28.   Classical attractors: point, circle, and torus.

we put together the four separate circles corresponding to the four periodic motions, getting a 4-dimensional torus as the attractor.

These are the three familiar types of behaviour recognized in classical work on differential equations, though of course their discoverers didn't think in terms of topological attractors in phase space. Each is deterministic, in the sense that the corresponding equations have unique solutions for unique initial conditions; and also predictable in the sense that, if you know the initial conditions accurately enough, the system will always stay close to what it would have done if those initial conditions were infinitely accurate. It's never possible in practice to measure anything exactly: there are bound to be limitations on the precision of measurement. So practical predictions have to be able to work with slightly inaccurate data; and the hope is that the predictions will also be only slightly inaccurate. For the classical attractors, this is so.

Strange attractors

So what the topologists asked themselves was: is every typical attractor classical? The restriction to typical cases occurs for the same reasons as in catastrophe theory (which pinched the idea from topological dynamics anyway): there are millions of infinitely rare accidents that complicate the picture impossibly if you try to describe them too, but which only occur in practice if some special constraint is forcing them to. In which case you'd be better off with a specially tailored theory to suit that constraint.

There are two main kinds of dynamical system, known as *discrete* and *continuous*. The difference is that in the first, time flows in definite steps 0, 1, 2, 3, . . . , whereas in the second it flows continuously. The discrete case is mathematically more tractable, especially for constructing and analysing fancy examples. And you can turn any discrete system into a continuous one, by 'interpolating' the behaviour between successive time instants. All that happens is that the phase space grows one extra dimension to give the interpolation somewhere to live. The technical name for the process is *suspension*. Conversely, you can often understand the behaviour of a continuous system by watching how representative points flow through some lower dimensional cross-section. This forms a discrete system since there is a definite interval between returns to the chosen slice. The upshot of all this is that for theoretical purposes 2-dimensional discrete systems are much like 3-dimensional continuous ones, 3-dimensional discrete like 4-dimensional continuous, and so on. I can now describe a 3-dimensional discrete system having an attractor that is typical, and not classical. Its suspension will be a 4-dimensional continuous system with the same properties. The phase space is a *solid* torus. At successive instants of time, its points move as follows. Stretch the torus out to ten times its length and one tenth the thickness, and wrap it round *ten times* inside itself, in a tenfold loop. (Actually, two will do in place of ten, but it complicates life later.) At the next instant it will stretch more, shrink more, and wrap a hundred times; then a thousand, ten thousand . . . Any initial point will move into successive tori in this sequence, so the attractor is the set of points common to all of the multiply wrapped subtori. It's pretty clear that this set is not a classical point, circle, or torus. The topologists call it a *strange attractor*. It is strange because, unlike the classical attractors, it has 'structure on all scales'. If you magnify a segment of a circle, it looks pretty much flat; and the same goes for a torus. But if you magnify the strange attractor, it retains its detailed structure of nested tori. A set with structure on all scales is called a *fractal*. So the distinction between the newfangled strange attractors and the familiar classical ones is geometric: the old ones are manifolds, the new ones fractals. The term *fractal attractor* is gaining popularity. I'll have more to say about fractals in the next chapter.

Let your digits do the walking

To see where the element of unpredictability comes in, consider an arbitrary initial point, and ask: where does it go? You get a solid torus by pushing a disc round a circle, which forms a 'core'. You can specify a point in a solid torus using three coordinates—two to specify a point in the original disc, and one for the angle the disc has to be pushed through. To keep the discussion simple, consider just this angle θ. Since we stretch the torus to ten times its length and wrap ten times, each step takes θ to 10θ.

It's convenient to measure θ so that the full 360° circle corresponds to one unit. We can tell roughly where a point is by dividing the circle into ten sectors labelled $0, 1, \ldots, 9$. Take an arbitrary starting-point, say at angle 0.37528 units. On the first wrap it goes to ten times this, namely 3.7528; but angle 1 is the full 360° and this is the same as angle 0. So we have to chop off the part in front of the decimal point, getting 0.7528. By similar reasoning this goes in turn to 0.528, 0.28, 0.8, and then settles at 0. This is striking. You can predict the entire future of the point by using the successive decimal digits of the angle representing it. At the nth step, it ends up in the sector indicated by the nth decimal digit.

This particular initial point eventually reaches 0 and stops, but that happens only because its decimal expansion terminates. Out to infinity it is really 0.375280000000 .... If instead we use 0.37528000 . . . 0063291 (say), with twenty-nine billion zeros where the dots are, its history is quite different. It also goes to sectors 3, 7, 5, 2, 8, and then sits in sector 0. But instead of staying there forever, as 0.37528 did, it only sits there for the next twenty-nine billion steps. On the twenty-nine billion and first, it hops to sector 6, then 3, then 2, then 9, then 1. And by extending the sequence of digits you can make it do whatever you wish. The initial points coincide for twenty-nine billion and five decimal places (far more accurate than any practical measurement could possibly be); but their eventual histories are *completely independent of each other*. This effect is called 'sensitivity to initial conditions'. Now in any interval, however small, you can find real numbers whose sequence of decimal digits, from some point on, follows any pattern you wish—including completely random series of digits. In fact *most* sequences of digits are random, see chapter 22. So as close as you wish to any initial condition you can find initial conditions leading to completely random motion.

Fractal weather

If strange attractors are typical, how come nobody spotted them earlier? There are two reasons. One is the classical obsession with analytical or closed-form solutions: these are too nicely behaved to lead to strange attractors. Numerical solution on a computer, only recently a viable option, produces strange attractors just as easily as classical ones. And in fact people *did* observe strange attractors in numerical studies. But, with very few exceptions, they assumed something had gone wrong, either with the theory, the calculation, or the experiment; so they threw the results away. Publication was never even contemplated. Random behaviour in a deterministic system was not expected, therefore it was not believed when it was seen. Only after the topologists showed that it *was* expected did people start to publish papers saying that they'd found it.

A gallant exception was a paper by the meteorologist E. N. Lorenz in 1963. Appropriately, Lorenz was modelling . . . the weather. However, his equations are a very rough-cut approximation to anything happening in real weather. The Lorenz system is an oscillator: it wobbles to and fro like a pendulum. But, quite unlike a pendulum, it doesn't oscillate regularly. In fact, its wobbles look pretty much random. The Lorenz system has a strange attractor.

Lorenz did publish his results, but (strangely!) they attracted little attention. So, when the topologists were developing their theories, they didn't know what Lorenz had done, and they rediscovered strange attractors in their own way. Only later was everything put together. And because the topologists hadn't known of the Lorenz attractor when they were setting up their general theory, it turned out to be an even more complicated attractor than the ones they could comfortably handle. In fact, a rigorous proof that it *is* a strange attractor has not yet been published, although the experts have a good idea how to go about finding one.

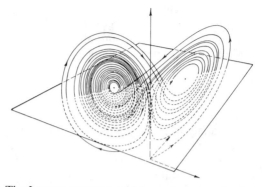

Fig. 29.   The Lorenz attractor, which produces 'random' dynamics in
a deterministic equation derived from a model of the weather. (From
John Guckenheimer and Philip Holmes, *Nonlinear Oscillations,
Dynamical Systems, and Bifurcations of Vector Fields*, Springer, New
York, 1983.)

Three theories of turbulence

One of the most classical parts of applied mathematics is the
dynamics of fluids. These are governed by various differential
equations, used for different purposes, and among the most
general are the *Navier-Stokes Equations*. These have an excellent track record of agreement with experiment, and apply
equally well to fluid flowing through channels and pipes, round
obstacles, and over dams. They describe the motion of waves;
and according to recent computer studies they even describe
how waves break. Despite these successes, there is a type of
fluid flow that has baffled hydrodynamicists for a century or
more: *turbulence*. The theories work well on smooth flows,
such as a tap turned on gently; but they have little to say about
the irregular frothing flow that occurs when a tap is turned on
full. On the whole, a fluid becomes turbulent when its speed
becomes too great. Turbulence is both regular and irregular.
In laboratory experiments, it usually sets in at much the same
point in the experiment, in much the same way, and with
much the same general appearance. It may appear suddenly or
gradually, or take on distinct forms, but it is repeatable. On
the other hand the actual motion is chaotic, random, and
unpredictable. Its statistical features do tend to have regular

patterns, and engineers often model turbulence with random terms in the equations. Even though a great deal of work has been done on turbulence, its fundamental *mathematical* nature is still a puzzle.

The Leray Theory of turbulence, dating from 1934, asserts that when turbulence occurs the Navier-Stokes equations break down as an adequate description of the fluid. In other words, turbulence is a fundamentally different phenomenon from smooth flow. The Hopf-Landau theory of 1948 disagrees, and holds that turbulence is a quasiperiodic flow in which infinitely many different period motions combine to give the appearance of randomness. In 1972 David Ruelle and Floris Takens criticized the Hopf-Landau theory on the grounds that it is topologically unlikely. It requires the fluid to undergo a series of changes which from the topological view point should never be able to get off the ground. An infinite series of coincidences is required for the Hopf-Landau mechanism to work. Ruelle and Takens suggested instead that turbulence is the formation of a strange attractor in the dynamics of the fluid. This is with hindsight a plausible idea: random behaviour and sensitivity to initial conditions are characteristic of turbulence and strange attractors alike. The physicist Mitchell Feigenbaum had a similar idea with a rather different topological scenario. The proposal was highly controversial and did not appeal to many classical hydrodynamicists.

How do you find out what a fluid is doing? The classical method is to eject a stream of dye, or insert probes or hot wires at various points. The development of the laser makes it possible to obtain much better results, more quickly, and with less disturbance to the fluid. The idea is to suspend tiny particles in the fluid, and bounce a laser beam off them. Their speed can be found using the Doppler effect, a shift in the wavelength of the reflected light. Experiments by Harry Swinney and his collaborators, using laser Doppler velocimetry, have done much to confirm that, at least in special laboratory systems, the onset of turbulence conforms to a strange attractor theory. The experiments yield precise measurements of how the fluid velocity at a single point varies over time. But how can we get from the velocity at a point to the dynamics of the entire fluid? Quantitatively, there's not much chance. But qualitatively, it's not hard. New techniques, known as *phase space reconstruction*, make it possible to deduce the topology of the corresponding attractor. These techniques are currently being developed into entirely new tools for analysing experimental data and extracting patterns.

What use is chaos?

A fair question. There are several ways to answer it.

The simplest is the 'what use is a new-born boby?' approach. New ideas have to mature before they affect company balance sheets.

Another is to compare it with the question 'what use are shock-waves?' The mathematics of compressible fluid flow is enormously complicated by their presence. They are the reason behind the 'sound barrier', which caused severe problems in the development of supersonic flight. You don't *use* shock-waves so much as overcome them. The point is, they're *there*. The existence of shock-waves is a mathematical discovery; but you don't make aerodynamics work any better by 'undiscovering' it. The shock-waves don't go away if you avoid the mathematics that proves their existence. In the same manner, chaos doesn't go away if you ignore it. It is a phenomenon that affects the whole of nonlinear dynamics, and hence almost all areas of applied science. Even if your only interest in chaos is to get rid of it, you'd better understand how it comes about. *Know your enemy.*

But the best answer is to tackle the question head on. Chaos *is* useful. It helps us to understand complicated phenomena, such as epidemics, or irregularities of the heartbeat. More than that: you can exploit chaos to attain a desired end. Let me offer three examples—the third one rather speculative, to stimulate your imagination rather than to prove any particular debating point.

The first example is mixing. Many industrial processes require ingredients to be mixed together thoroughly and uniformly. However, you don't want your mixing equipment to be incredibly complicated and 'random': you want something nice and simple that your engineers can maintain in good condition. In other words, you want a simple deterministic system that produces random behaviour. Mixing-machines are chaos generators, virtually by definition. Think of an egg-beater, for instance. Two interlocking paddles rotate at constant speed: the egg goes all over the place. A deterministic process produces a chaotic result.

Chaos has led to new insights into the mathematics of mixing. In particular it turns out that many simple mixing devices tend to leave 'dead areas' that are not well mixed. The reason is that the materials being mixed are (approximately) incompressible, and the dynamics is rather special: it is an instance of what's called a *Hamiltonian system*. The dynamics of the solar system is Hamiltonian; and we've already seen how counter-intuitive some features of celestial mechanics can be. Hamiltonian chaos has a very special flavour. Uniformly chaotic flow is hard to achieve in Hamiltonian systems: one of the basic phenomena is the tendency for 'islands of stability' to occur, even when surrounded by a sea of chaos. These islands create the 'dead areas' of poor mixing.

A second application is to noise reduction. A typical problem is to extract meaningful conversation from a recording made in a room with a *very* noisy air-conditioner. The traditional approach is to think of the conversation as a 'signal' and the air-conditioner as a source of 'noise'. The noise is treated as a random disturbance to a regular signal. Provided the signal-to-noise ratio is large enough, much of the noise can be filtered out by appropriate mathematical procedures. But when the air-conditioner noise exceeds the level of conversation by a large amount, there's no hope for the traditional method.

Let's think chaos, however. From this viewpoint, it is the *air-conditioner* that is the signal—it is produced by a deterministic mechanical process. The conversation is non-deterministic—you can't hope to predict what the people will say—and should be thought of as *noise*. Our perception of what is noise and what is signal is now reversed. Instead of the noise being much larger than the signal, the signal is much larger than the noise. Admittedly, it's the wrong sort of signal for the traditional approach—it's not regular enough. The new methods of phase space reconstruction, however, are designed to handle chaos, and they let you make short-term predictions of the chaotic signal, which filter away the noise. Great: now you have a cleaned-up recording of an air-conditioner, with the conversation removed. The final step is to subtract this from the original recording, air-conditioner plus conversation. The air-conditioner sounds cancel out, leaving only the conversation. This method works: the conversation is 95 per cent intelligible even when the air-conditioner is 25 decibels louder.

These are technological uses of chaos. What about 'natural' uses? Surely living organisms have no use for chaos?

Maybe they do. The most obvious example, our third application of chaos, is flight from a predator. The motions of the prey should be unpredictable, to confuse the predator. But they have to be generated by the prey's nervous system. The simplest way to generate unpredictable movement is to put some neural circuit into a chaotic state, and let that take charge. Some houseflies have a habit of flying along paths that look like random polygons, with sudden changes of direction at irregular intervals. Presumably they're doing this to confuse predators, such as humans armed with fly-swatters. But these 'random' changes have to be controlled by the fly's nervous system, which is not the most complicated of computational devices. Low-dimensional chaos would seem an attractive option. I know of no experimental data on this particular form of behaviour, but it would be interesting to take a look.

If prey do indeed use chaos to help them escape, then it becomes evolutionarily worthwhile for predators to evolve neural circuitry for some type of 'phase space reconstruction', to predict where the prey will go next. This sets up suitable conditions for what Richard Dawkins calls an evolutionary 'arms race'. The result should be increasingly sophisticated chaos-generators from the prey, and better phase space reconstruction from the predators. But perhaps the speculation should stop here.

Symmetric chaos

Traditional dynamics deals with regular patterns, whereas chaos is the science of the irregular. Its aim is to extract new kinds of pattern from apparent randomness, pattern that would not be recognized as such in a traditional approach. This description makes it tempting to think that order and chaos are two opposite polarities. However, some fascinating computer experiments have revealed that some systems manage to combine order and chaos at the same time.

The traditional source of pattern is symmetry. Anything with a high degree of symmetry is automatically patterned. I've already mentioned the interaction of symmetry and dynamics: symmetry-breaking for steady states and (via Hopf's wobble) periodic states. A few years ago, Chossat and Golubitsky asked themselves what would happen if you took a symmetric dynamical system and made it operate in the chaotic regime. The answer is that you get strange attractors—with symmetry.

The method is simple. Choose a group of symmetries, say transformations of the plane, such as the symmetry group of a regular polygon. Invent a mapping from the plane to the plane that 'preserves' these symmetries. This means that if you start with two symmetrically related points and apply the mapping to them both, then the two resulting points must be related by the same symmetry. For example, suppose one symmetry is rotation through 72° (five-fold symmetry). Take a point A, and rotate it through 72° to get another point B. Suppose the mapping sends A to A' and B to B'. Then B' must be A', rotated through 72°.

For any given group of symmetries, it is just a technical calculation to find such symmetry-preserving transformations. Usually there are huge numbers of them, with lots of adjustable coefficients—numbers that you can choose at will. Having done so, you just iterate the mapping millions of times. That is, you choose a point, apply the mapping, and repeat. The resulting set of points (with the first few hundred omitted to remove 'transients') is the attractor for your chosen mapping.

You get a different attractor for each choice of coefficients. Most of of them are symmetric. Many are also chaotic. Different choices of symmetry group yield different symmetries in the attractors. There are 'icons' with the symmetries of a starfish. Indeed, one closely resembles the creature known as a sand-dollar, a fact that is presumably a visual pun. And a different choice of symmetry group produces a chaotic 'quilt' design with a pattern that repeats in a lattice. Patterns of this kind have been investigated by Golubitsky and Mike Field.

There's more to this idea than just pretty computer pictures.

Fig. 30. Symmetric chaos. A 'quilt' attractor for a system with
lattice symmetry. (Mike Field and Martin Golubitsky)

Symmetry is fundamental in nature. So, we now realize, is
chaos. Therefore the two in combination should be especially
interesting. Symmetric chaos is a new area of research, but
it has already progressed beyond computer experiments to
proving theorems. It appears to offer an explanation of some
puzzling real-world phenomena, notably the occurrence of
'patterned turbulence'. For example, the Taylor–Couette
system can produce a state known as turbulent Taylor vortices:
the usual layered vortex structure, but with a superimposed
fine texture of turbulence. This appears to be a physical realiz-
ation of a symmetric chaotic attractor.

  These attractors also raise interesting questions about the
nature of complexity. The amount of 'information' required to
*de*scribe the points in a chaotic attractor is huge. The attractor
looks very complicated. But very little information—a single

formula plus a few numerical coefficients—is needed to *pre*scribe the mapping that generates the attractor. You could transmit the formula and coefficients over a phone line in a few seconds. To transmit the end result, coordinate by co-ordinate, would take hours.

This implies that the visual simplicity or complexity of an object may be misleading. What is more important, but far deeper, is the complexity of the *process* that generates it. This in turn means that at least some of the enormous complexity that we see in the world of living creatures may have simple causes. The mathematical development of this point of view is called algorithmic information theory: see chapter 21.

When Einstein's Theory of Relativity achieved public recognition, most people intepreted it as saying that 'every-thing is relative', a comfortable philosophy that, for example, justifies the rich ignoring the poor on the grounds that others are yet poorer. However, that's *not* what Einstein was saying: he was telling us that the speed of light is *not* relative, but absolute. It should have been named the Theory of Non-relativity. Something similar is happening with Chaos Theory. Some people are taking it to mean that 'everything is random', and using that to justify economic or ecological mismanage-ment. How unfair to blame the Chancellor of the Exchequer for not controlling inflation or unemployment, when 'every-body knows' these are chaotic! But Chaos Theory offers no such excuse. Its message is that *some* things that we think we understand may behave in very funny ways, *some* things that appear random may obey laws we haven't yet spotted, and *most* things don't fit into any of these categories at all. Indeed Chaos Theory has opened up new methods for *controlling* systems that appear to behave randomly: the Chancellor must shoulder more, not less, of the blame when the economy goes awry.

Rich mixture

Fifty years ago the difference between determinism and ran-domness was straightforward. Equations without random terms were deterministic, and that meant predictable. Which in turn meant that if you were observing something that seemed to behave randomly, then you needed a random model. But this black-and-white picture has given way to a complete spectrum, with varying shades of grey. A system can be deterministic but unpredictable, have no random terms but behave randomly. Many irregular phenomena may perhaps be modelled by simple deterministic equations. On the other hand, simple equations are no longer assumed to imply simple behaviour. Examples other than turbulence include the Earth's magnetic field, with its random reversals; electronic oscillators; the growth of insect populations; and properties of liquid helium.

A typical dynamical system is usually neither wholly random, nor wholly predictable. Its phase space has regions corresponding to each, often intertwined to form an intricate and rich mixture. And while we now understand that this is so, we can only guess where this observation may lead. It will certainly require new kinds of experimental method; one is described in the next chapter. It may well alter the accepted view of experimental 'predictability' as a criterion for acceptance of a theory.

18

The two-and-a-halfth dimension

Clouds are not spheres, mountains are not
cones, coastlines are not circles, and bark is not
smooth, nor does lightning travel in a straight
line.

Benoit Mandelbrot

How long is the coast of Britain? It depends on how carefully
you measure it. There are large wiggles (the Wash and the
Bristol Channel), medium-sized wiggles (Flamborough Head,
Orford Ness), small wiggles (gaps between rocks on the fore-
shore at Dover), diminutive wiggles (irregularities in a single
rock) . . . The closer you look, the more wiggles you find.
So the more carefully you measure the coast, the longer it
appears to become. What is the area of the human lung? The
same problem arises, because the surface of the lung is folded
and branched in a very complicated way, in order to squeeze a
large oxygen-gathering capacity inside a small human chest.

The traditional models of theoretical science are objects
such as curves and surfaces. Usually these are assumed to be
smooth—so that calculus can be used to study them—which
means that on a large enough scale they appear flat. For
example the Earth, on a rough approximation, is a sphere; and
to its inhabitants it appears flat, because not enough of it is
visible to notice the very gentle curvature. But the coastline of
Britain and the surface of the lung do not look flat, even under
a powerful microscope. They remain crinkled. Such objects,
possessing detailed structure on many scales of magnification,
abound in nature; but only recently have they been identified
by scientists and mathematicians as something worthy of
study in its own right. The result is a new breed of geometric
figure, called a *fractal*. The name, derived from the Latin
for 'irregular', was coined by Benoit Mandelbrot, an IBM

Fellow at the Watson Research Centre in New York State. Mandelbrot has been the prime mover in the recognition of fractals as a new kind of mathematical object, well suited for the modelling of natural phenomena involving irregularity on many scales.

   We generally say that we live in a space having three dimensions (north, east, up). A plane has two dimensions, a line one, and a point zero. Mathematicians have devised spaces with any whole number of dimensions. It turns out that the sensible way to assign a dimension to a fractal leads to *fractional* values, such as 1.4427. This bizarre property is very basic and reflects the fractal's behaviour under changes of scale. The very *existence* of such forms suggests an enormous range of new physical and mathematical questions, by directing our attention away from the classical obsession with smoothness. What happens when light rays pass through a medium whose refractive index varies fractally? If they reflect off a fractal mirror? Or, for the more practically (?) minded: how will the passage of a laser beam through a turbulent atmosphere affect its ability to destroy a nuclear missile? What happens to a radar beam when it bounces off an irregular landscape? What sort of noise will a drum with a fractal rim make? Once our eyes have been opened to the distinctive character of fractals, we observe that the world is full of them. As yet, we know very little about their deeper nature.

Gallery of monsters

Around the turn of the century a number of mathematicians introduced some novel geometrical forms: highly 'irregular' curves and surfaces. They were seen as 'pathological' objects whose purpose was to exhibit the *limitations* of classical analysis. They were counterexamples serving to remind us that mathematics's capacity for nastiness is unbounded. Curves that fill an entire square, curves that cross themselves at every point, curves of infinite length enclosing a finite area, curves with no length at all. For instance, during the eighteenth and early nineteenth centuries it was assumed that a continuous function has a well-defined tangent (that is, is differentiable in the sense of calculus) at 'almost all' points. In a lecture to the Berlin Academy in 1872, Weierstrass showed that this is untrue, exhibiting a class of functions which are continuous everywhere but differentiable nowhere. The graph of the Weierstrass function has no gaps, but is so irregular that it has no tangent at any point. There are other continuous but nowhere differentiable functions, including one creature aptly called the *blancmange function*, its graph having multiple peaks.

A classic pathological case is the *Cantor set*, obained from an interval by throwing most of it away. Delete the middle third, to leave two smaller intervals; delete the middle third of each of these, and repeat forever. The total length removed is equal to the length of the original interval; yet an uncountable infinity of points remain. Cantor used this set in 1883; it was known to Henry Smith in 1875. In 1890, Peano constructed a curve passing through *every* point of the unit square. This demolished the current definition of the 'dimension' of a set as the number of variables required to specify its points. The *Peano curve* lets us define the points in the two-dimensional square by using a single variable: how far along the curve we

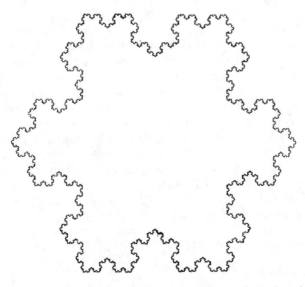

Fig. 31. The von Koch snowflake curve has infinite length and encloses a finite area. It was one of the earliest fractals to be invented.

must travel to reach the chosen point. In 1906 Helge Von Koch found a curve of infinite length containing a finite area: the *snowflake*. It is obtained from an equilateral triangle by erecting on each side a triangle one-third as large, and repeating this process indefinitely. Like Weierstrass's curve, the snowflake is continuous but has no tangents. However, it is not the graph of a function.

Most mathematicians of the period accepted that these strange sets did demonstrate that there were limits to the applicability of classical analysis. But they were perfectly prepared to operate *within* those limits, and few saw any reason to study the 'pathologies' in their own right. They were seen as artificial objects, unlikely to be of any importance in science or mathematics. Poincaré called them a 'gallery of monsters'. Hermite wrote of 'turning away in fear and horror from this lamentable plague of functions with no derivatives', and he tried to prevent Lebesgue from publishing a paper on non-differentiable surfaces. As we saw in chapter 14, Lebesgue invented the enormously important method of integration that still bears his name; he was no intellectual lightweight. He later wrote of Hermite's rejection of his paper: 'He must have thought that those who make themselves dull in this study are wasting their time instead of devoting it to useful research.' And he also said: 'To many mathematicians I became the man of the functions without derivatives. And since the fear and horror which Hermite showed was felt by almost everybody, whenever I tried to take part in a mathematical discussion there would be an analyst who would say, "This won't interest you; we are discussing functions having derivatives."' Lebesgue's experience shows that we must be careful to distinguish between the mathematical mainstream and the purely conventional or fashionable.

To be fair, it is clear that the undue proliferation of such sets, without any clear purpose, is a pretty futile exercise. But with hindsight, we observe mathematics starting to grapple with the structure of fractals.

Topological dimension

There are many different ways to define the dimension of a space. If it is a smooth manifold (the multidimensional analogue of a surface) then the dimension is the number of variables in any *smooth* coordinate system. (Requiring smoothness excludes the Peano curve but restricts the range of spaces to which the definition applies.) For example the surface of a sphere is 2-dimensional, because the two variables latitude and longitude provide a coordinate system. Obviously the dimension in this sense must be a positive integer. Engineers will recognize it as the number of 'degrees of freedom'.

Poincaré generalized this definition to arbitrary topological spaces by insisting that:

(a) The empty set has dimension -1.
(b) If the boundaries of small neighbourhoods of all points in the space are $(n-1)$-dimensional, then the space is n-dimensional.

This is called an *inductive* definition: its defines 0-dimensional spaces in terms of -1-dimensional spaces, then 1-dimensional in terms of 0-dimensional, and so on. Let us call the result the *topological dimension*. By its nature it is a topological invariant, and again must be a positive integer.

In the theory of fractals, something stronger than topology is required. A snowflake curve is topologically the same as a circle! It turns out that the *metric* structure—notions of distance—must also be taken into account. It is necessary to seek a definition of dimension that extends the usual one for manifolds, but reflects metric, not topological, properties.

Fractal dimension

Take a piece of string. Take another the same length, and place them end to end: the result is, of course, twice the size. Starting with a square of paper, however, you need *four* copies to make a square twice the size. With a cube of cheese, it takes *eight* copies to double the size. If four-dimensional cheese existed, you would need *sixteen* copies. In other words, to double a d-dimensional hypercube you need $c = 2^d$ copies. Therefore the dimension is given by solving for d:

$$d = \log c / \log 2.$$

For example, if we can find an object whose size doubles if three copies are stuck together, then it would have dimension

$$d = \log 3/\log 2 = 1.4427\ldots$$

More generally, to multiply the size of a d-dimensional hypercube by a factor a you need $c = a^d$ copies, and now

$$d = \log c/\log a.$$

It is easy to see that (one side of) the snowflake curve is made up of four copies of itself, each one third the size. So $a = 3$, $c = 4$, and

$$d = \log 4/\log 3 = 1.2618\ldots$$

The dimension thus defined may appear a trifle unorthodox (like the blancmange function—joke) but it has a serious purpose: it reflects the *scaling* properties of the curve. And that's what makes the *similarity dimension* that we have defined so basic. A set to which it can be applied is said to be *self-similar*. The idea can be modified to apply to other sets too; the final version is called Hausdorff–Besicovitch dimension, or, more simply, *fractal dimension*.

In 1961 Lewis Richardson took a look at the coastline-measuring problem for a number of countries. Take a ruler of fixed length e and move along the coast in steps of size e: add the results to get a total length $L(e)$. Because shorter rulers detect smaller wiggles, $L(e)$ must get bigger as e gets smaller. Richardson found experimentally that there is a number D such that $L(e)$ behaves like e^{1-D}. For Britain, D is about 1.25. The same value holds for Australia; for the land-frontiers of Germany and Portugal it is closer to 1.15; for South Africa roughly 1. By looking at the snowflake's length, measured the same way, we see that D is just the fractal dimension. So Britain's coastline is a fractal curve of dimension about 1.25. For some reason South Africa is much smoother. Indeed you can think of D as a measure of the degree of roughness. The snowflake curve is pretty much as rough as the coast of Britain. It won't make a very good model of a coastline, though, because it has much too definite a pattern. (It's not really a good idea to describe a fractal as being 'irregular'; 'rough' is better.) There is an idea of statistical self-similarity that can be applied to random curves; and a random fractal of dimension 1.25 looks just like a piece of coastline. The self-similarity means that if you take a section of coastline and magnify it, then the result is equally plausible as a stretch of coastline. To put it another way, if you are given a map of the coastline with no scale marked, you won't be able to work out the scale by analysing the map.

Nature's fractals

Structures resembling fractals are common in Nature. Of course, a natural object won't have structure on *all* scales, from the infinitely large to the infinitely small—or at least, if it does, we can never observe all of it—so a fractal model is only an approximation. But when an object possesses detailed structure on a wide range of scales, then a fractal may be a better model than a smooth curve or surface.

William Harvey, discoverer of the circulation of the blood, noted that close to almost all points of the body there will be found both an artery and a vein. Modelling this by a logical extreme, we can insist that there should be both an artery and a vein infinitely near every point. Or, in different words, every point of the body not already inside a vein or artery should lie on the boundary between the two blood networks. In addition, for efficiency the total volume of arteries and veins should be small compared to that of the body (or else living creatures would be merely large bags full of blood). Mandelbrot puts it this way: 'From a Euclidean viewpoint, our criteria involve an exquisite anomaly. A shape must be topologically two-dimensional, because it forms the common boundary to two shapes that are topologically three-dimensional, but it is required to have a volume that not only is non-negligible compared to the volumes of the shapes that it bounds, but is much larger!' A simple fractal construction resolves the paradox. All criteria can be met provided the blood networks branch infinitely often, with rapidly decreasing diameters for the blood vessels at each stage. Body tissue then becomes a fractal surface; it has topological dimension 3, but fractal dimension 2. We are not walking blood-bags—we are fractals incarnate! Fanciful though this idea may seem, it has a compelling intuitive quality for anyone familiar with fractals. The paradoxical properties of fractal geometry can liberate our intuition and save us from unwarranted assumptions.

For example, consider Saturn's rings. Until recently these were thought of as something akin to flat discs of matter. Telescopic observations suggested the presence of five or six annular zones, separated by clearly defined gaps (supposedly due to, and explained by, gravitational resonances with some of Saturn's moons). When the two Voyager spacecraft actually went to Saturn and took a look, they found something more complicated: a fine structure of thousands if not millions of rings. New gaps in the rings failed to fit the 'resonance' theories. Saturn's rings look much more like a fractal than a flat disc. Now the gravitational properties of fractally distributed matter have not been studied; but they are unlikely to be exactly the same as those of flat discs. It's an uncharted area of research; and it may be that the fractal model could explain the failure of the resonance calculations. Simplifying hypotheses such as smoothness are not always appropriate; and an awareness of the fractal alternative could help researchers from making such an assumption in the wrong context.

Most plants have a fractal structure. Just think of a tree with its ever-decreasing hierarchy of limbs: trunk, bough, branch, twig . . . and the bog down in the valley-oh! Imagine an insect living on a leaf with a fractal surface. The smaller the insect, the more tiny dents it will be able to get into. So the effective surface area of the leaf is much larger for smaller insects, just as the coastline of Britain seems longer if you use a finer ruler. New 'ecological niches' arise for ever tinier creatures. Experiments conducted in 1985 by D. R. Morse, J. H. Lawton, M. M. Dodson, and M. H. Williamson have confirmed this relation between insect size, insect population, and fractal dimension of the host plant. The fractal nature of plants may have influenced the evolution of insects.

Brownian motion

Another example of fractals in Nature is Brownian motion, mentioned in chapter 14. Small particles suspended in a fluid undergo rapid irregular movements, caused by thermal agitation of molecules in the fluid. The molecules bump into the particles, and give them a shove. In 1906 Jean Perrin wrote:

> The direction of the straight line joining the positions occupied at two instants very close in time is found to vary absolutely irregularly as the time between the two instants is decreased. An unprejudiced observer would therefore conclude that he is dealing with a function without derivative, instead of a curve to which a tangent should be drawn. At certain scales and for certain methods of investigation, many phenomena may be represented by regular continuous functions. If, to go further, we attribute to matter the infinitely granular structure that is in the spirit of atomic theory, our power to apply the rigorous mathematical concept of continuity will greatly decrease.

Perrin went on to discuss the idea of density of a gas as a limiting value of the mass-to-volume ratio for a small region. He pointed out that this concept, the basis of the continuum mechanics of gases, makes no physical sense in an atomic theory, where the limiting value is 0 except at isolated points (the atoms) where it becomes infinite. In the atomic regime, fluctuations dominate. Perrin realized that the 'pathological' continuous but non-differentiable curves of Weierstrass and von Koch occur in Nature, in processes as fundamental as molecular motion. In the 1930s, Norbert Wiener formulated a mathematical model of Brownian motion possessing exactly this feature of non-differentiability of the paths of particles.

Fractals in the mainstream

In mathematics itself, fractals passed from fashion to obscurity, and are now emerging from obscurity to take their rightful place in the mainstream. The strange attractors of chaotic dynamics are fractals; and there are other places where fractals are contributing new insights. One area, rather closely related to dynamics, concerns the behaviour of a function under iteration. That is, we take a function f and a number x, and form in turn the sequence $x, f(x), f(f(x)), f(f(f(x))), \ldots$ by repeatedly applying f. Let $f^n(x)$ be the nth term in this sequence—the nth *iterate* of x under f. For example suppose that $f(x) = 2x$, and we start at $x = 1$. Then $f(1) = 2, f(f(1)) = 4, f(f(f(1))) = 8$; and in general $f^n(1) = 2^n$. For more complicated functions f it is not possible to write down a general formula, but it is still possible to study the way the sequence behaves. An important example is the case $f(x) = ax(1 - x)$

where x lies between 0 and 1, and a is a parameter between 0 and 4. The long-term behaviour of the iterates of this function depends on the value of a. For $a < 3$ they converge to a single value; for a slightly greater than 3 they oscillate between two values, then four, then eight . . . and at a value of a near 3.7 the system goes chaotic.

Mandelbrot set

One particular fractal has attracted enormous public attention. It has become virtually synonymous with 'Chaos Theory'—the popular conception of chaotic dynamics and fractals. A group of hoaxers even created a 'corn circle' in its image in the Cambridge countryside. (We know it must have been hoaxers for a variety of reasons: mostly, they got the shape *wrong*). It was invented by Mandelbrot, and has become known as the Mandelbrot set.

The Mandelbrot set is a test-bed for the theoretical investigation of fractal structures that come up in an area known as *complex dynamics*. This is not just dynamics with bells and whistles—though it often looks like it. It is the dynamics of mappings of the complex plane under iteration. The subject's 'boost phase' was the work of Pierre Fatou and Gaston Julia in the 1920s. The 'pre-launch phase' goes back at least as far as Gauss, who made some basic discoveries in his theory of the arithmetico-geometric mean. We are now in the flight phase.

Suppose that c is a complex constant, and consider the mapping $z \rightarrow z^2 + c$ where z is also complex. What happens to z? In detail, there are many possibilities; one of the simplest is that z disappears to infinity. The set of initial points z that do this is called the *basin of attraction* of the point at infinity; and its boundary is the *Julia set* corresponding to c. With modern computer graphics it is easy to draw Julia sets. They can take a remarkable variety of forms: circles, pinwheels, rabbits, sea-horses, starry skies. To bring some order, we ask only whether the Julia set is connected. Is it all in one piece, or does it fall apart into dust? Note that this is a question about the parameter c, not the variable z. The points c in the complex plane for which the Julia set is connected comprise the Mandelbrot set.

One method of drawing the Mandelbrot set, developed by

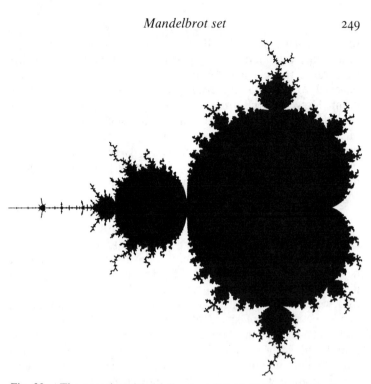

Fig. 32. The notorious Mandelbrot set (H.-O. Peitgen and D. Saupe (eds.), *The Science of Fractal Images*).

John Hubbard and exploited to great effect by Heinz-Otto Peitgen and Peter Richter, involves plotting colour-coded contours that home in on its intricate boundary. It is these pictures that feature on the postcards, posters, and coffee-mugs of shops such as Kensington's *Strange Attractions*. If you take any region near the boundary of the Mandelbrot set and zoom in on it, you see an ever-changing scene of complex geometric structures. If you look in just the right place, you see tiny Mandelbrot sets, complete in every detail—though each has its own characteristic external decoration. If you zoom in towards the point c, then the picture begins to re-semble, ever more closely, the Julia set for c. In short, the Mandelbrot set is all possible Julia sets, glued together.

Because of this complexity and depth of detail, the

Mandelbrot set has been described as 'the most complicated object in mathematics'. It is perhaps worth explaining in just what sense this clichéd statement is true. It is extremely complicated to *describe* the Mandelbrot set in detail, point by point. However, the *process* that generates it ($z \rightarrow z^2 + c$) is remarkably simple. So the Mandelbrot set isn't *really* complicated: it just looks that way. This is precisely the distinction that I made in the previous chapter, between description and prescription, process and result. There are plenty of things in mathematics that are much, much more complicated than the Mandelbrot set.

It is, however, complicated enough to pose some really tricky problems for mathematicians. One basic one is to prove that it really is a fractal. Yes, it certainly looks like one; but can you prove it? In 1991 Mitsuhiro Shishikura did so. Indeed, he proved that the boundary of the Mandelbrot set has fractal dimension 2, the largest possible value. This is the same value that would be attained by a space-filling curve—a curve that fills some region of space of non-zero area—and raises the question whether the boundary of the Mandelbrot set might actually be such a curve. However, different curves can have equal fractal dimensions, so this particular value of the dimension need not imply that the curve fills space. The result does mean, however, that the boundary of the Mandelbrot set is just about as wiggly as it is possible to be while living in a plane.

Recall that locally, near a point c, the Mandelbrot set looks like a Julia set for that value of c. Shishikura's proof relies on finding a sequence of Julia sets whose fractal dimensions approach 2 as the parameter c that defines them approaches the boundary of the Mandelbrot set. He doesn't define the sequence explicitly: he proves that it exists. His techniques are of considerable interest, and open up new possibilities in complex dynamics.

Digital sundial

Fractals have revitalized a classical area of mathematics known as geometric measure theory. This is where A. S. Besicovitch originally defined his notion of dimension, now renamed 'fractal dimension'. One problem in that area, solved by a fractal construction, asks what relation there is between the different projections—shadows—of a set in three-dimensional space. The answer, proved by Ken Falconer in 1986, is 'none whatsoever'. Roy Davies proved the same result for a restricted case, subsets of the plane, in 1952.

Imagine some set, possibly very complicated, hovering in space. On one side of it place a lamp (infinitely distant so that its rays are parallel) and on the other place a flat screen. You get a shadow. Move the lamp and screen round: you get another shadow. How does the shadow change? That the answer may be surprising is shown by an ancient puzzle: make a cork that will fit a round hole, a square one, or a triangular one. The shadow of the cork, from three distinct directions, has to be round, square, or triangular. It's not hard to concoct such a shape.

Falconer proves that, given any prescribed list of shadows corresponding to directions in space, there exists some set that produces 'almost' those shadows when illuminated from those directions. 'Almost' here means that the actual shadows may differ from those prescribed by a set of zero area. His construction is a kind of iterated venetian blind. A venetian blind lets light through in one direction, but not in others. What we

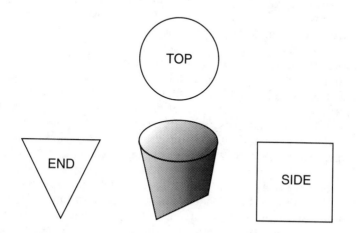

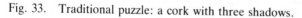

Fig. 33. Traditional puzzle: a cork with three shadows.

want is the opposite: something that casts a solid shadow in one direction, and nothing in others. To this end, we modify the venetian blind. If its slats are replaced by tiny venetian blinds pointing in some other direction, it now lets light through in two directions. By repeating this process you can get a set that lets light through in nearly all directions, with one exception. Cut such a blind to the shape of the shadow that you want to be produced in that direction. Now repeat with a new blind, a new direction, and a new shape. Combine all the resulting sets into one: it has arbitrarily prescribed shadows. The 'zero area' differences arise when you try to fill in the technical details in this approach. Falconer describes the ideas as the 'digital sundial'. You can make a shape which, when hung in sunlight, produces a moving shadow; and at each moment, the shape of the shadow tells you the time *in digits*. For example, at 11.17 the shadow will form the same shape as the digits 11.17.

Through fractal eyes

The contribution of fractals to our understanding of the natural world is not so much one of 'technology' as of what used to be called natural philosophy. Instead of being a computational tool, it is a clue towards an integrated world-view, an organizing principle. The recognition of fractals as basic geometric forms of a new kind, with their own mathematical structure, opens our eyes to a new range of phenomena and sensitizes us to new points of view.

At the moment, fractal models are descriptive rather than explanatory. We can *see* that fractal mountains of the right dimension look just like real mountains; but we lack any real idea of the reasons, in terms of erosion processes or fracture properties of materials, that are responsible for their fractal nature. We cannot yet compute even as basic an item as the fractal dimension of a model mountain from physical principles. Of course, a good descriptive framework is important. If cellular tissue is best modelled by a fractal, then there's not much point in a theory that treats the cell as a rectangular slab of homogeneous goo, or a living creature as a uniform sphere. What are the effects of fractal structure on a mathematical problem? When are they important and when do they make little difference? What are the causes of fractal forms? There are many such questions, but as yet few answers. They hint at a deeper structure of which fractal geometry is the outward sign.

Fractals are important, not because they have solved vast open problems in mathematics, or helped us to build a better mousetrap, or unlocked the secrets of rotating black holes. They are important because they suggest that, out there in the jungle of the unknown, is a whole new area of mathematics, directly relevant to the study of Nature. The great strength of mathematics is its ability to build up complex structures from a few simple key ideas. Once the skeleton of such a structure appears, each new piece can be attached in its rightful place. Without awareness of the skeleton, the pieces lie scattered and unappreciated. We now have the skeleton of a theory of fractals. The challenge to the mathematicians of the next century will be to mould the flesh on to those already fascinating bones.

19

The lonely wave

The physicists could but make the best of it, and
went around with woebegone faces sadly com-
plaining that on Mondays, Wednesdays and
Fridays they must look on light as a wave; on
Tuesdays, Thursdays, and Saturdays, as a particle. On Sundays they simply prayed.

Banesh Hoffmann

In 1834 the engineer and naval architect John Scott Russell
was riding along the side of the Edinburgh–Glasgow canal,
watching a boat being pulled along by two horses. 'The boat
suddenly stopped', he wrote in a report to the British Association in 1844.

Not so the mass of water in the channel which it had put in motion; it
accumulated around the prow of the vessel in a state of violent
agitation, then suddenly leaving it behind rolled forward with great
velocity, assuming the form of a large solitary elevation, a rounded,
smooth and well-defined heap of water, which continued its course
along the channel apparently without change of form or diminution of
speed. I followed it on horseback, and overtook it still rolling on at a
rate of some eight or nine miles an hour, preserving its original figure
some thirty feet long and a foot and a half in height. Its height
gradually diminished, and after a chase of one or two miles I lost it in
the windings of the channel. Such in the month of August 1834 was
my first chance interview with the singular and beautiful phenomenon
which I have called the Wave of Translation.

Russell's extraordinary observation lay dormant and neglected
for over a century, one of the numberless relics in science's
graveyard of undigested facts. But in 1965 it sprang to life,
revealing the existence of an entirely new mathematical object, the *soliton*, having applications throughout the physical sciences. It has led to the discovery of deep connections
between quantum field theory and those most abstract of

topics, algebraic topology and algebraic geometry. It may well provoke a revolution in quantum mechanics leading to a new understanding of the physics of the ultimate constituents of matter and the forces that shape our universe.

Lumps and ripples

Light, and other forms of electromagnetic radiation, make the universe what it is. But what *is* light? The ancient Greeks thought that it was a stream of particles, emitted by the eye of the observer. In 1680 Newton also thought it was a stream of particles, but emitted by the light source; bouncing off objects like balls on a snooker table, and producing the sense of vision when a particle chanced to enter the observer's eye. But in the rival theory of Huygens a decade later, light was not a particle, but a wave, travelling like ripples on a pond. At the time, most scientists ended up siding with Newton, although Newton himself performed experiments that are nowadays used to demonstrate that Huygens was right!

You can tell waves from particles by seeing what happens when they collide. Particles bounce, but waves *interfere*. If you drop two pebbles into a pond, the expanding rings of ripples do not bounce off each other. Instead they overlap, reinforcing each other where two wave-crests coincide, cancelling where a trough meets a crest, or producing a deeper depression when trough meets trough. Such interference patterns are characteristic of waves. In 1890 Thomas Young passed light through two parallel slits and observed interference. Light was a wave.

On the other hand, in 1922 Arthur Compton bombarded electrons with X-rays, and found that they interacted as if X-rays were particles. Which would have been fine, except that X-rays are just another form of electromagnetic radiation, effectively just light of very short wavelength. So is light a wave, or a particle? Well . . . yes! Of course, it could be worse. Electrons, at least, *are* particles, right? Wrong. In 1925 Louis de Broglie did the slit experiment with electrons, and got interference patterns again. He suggested that many objects thought to be particles might share this strange wave-particle duality, and he was soon proved correct. The universe is made not of waves and particles, but a host of wavicles, changing their disguise with the facility and enthusiasm of the Scarlet Pimpernel.

In today's quantum mechanics, a particle is described by a kind of probability wave, telling you how likely it is to find it in a given position. The wave is concentrated around the position you are likely to observe in an experiment, and follows this position without changing shape as the particle moves. The commonest waves in mathematics are of infinite extent, so the early theories used 'wave packets', superimposing the infinite ripples to produce a single localized lump. Unfortunately, these wave-packets don't hang together very effectively: they spread out as time passes. What is needed is a simple and natural theory in which non-dispersive localized waves can occur.

Waves that bounce

Russell did some experiments on his Wave of Translation, by dropping a weight into water at one end of a long channel. He found that the speed of the wave depends on how big it is—a result that does not hold for the classical linear theory of water waves. 'Linear' means that if you superimpose two solutions you get another solution. Most of the equations of mathematical physics at that time were linear, because nobody could solve anything else. Usually some approximation was used to obtain the linearity—shallow waves, low speeds, or the like. In 1871 J. Boussinesq approximated the problem by assuming the wave is much broader than the depth of water, and got a similar result for the speed. John William Strutt, Third Baron Rayleigh, had the same idea independently in 1876. Their equation was non-linear, but surprisingly they were able to solve it. In 1895 Diederik Korteweg and Hendrik de Vries found a similarly non-linear equation, the Korteweg–de Vries or KdV equation, and showed it has a solitary wave solution, where a single hump moves without changing shape. Again, the speed depends on the size of the hump, with bigger waves moving faster. The fact that these non-linear equations could be solved explicitly was seen as something of a coincidence. Conincidences don't lead to nice general theories—or so it was felt—and the topic was almost forgotten. Korteweg's several obituaries do not mention this now famous discovery, made while de Vries was writing a dissertation under his guidance. De Vries was a schoolteacher and very little is known about him. Mathematics continued blissfully along its linear course.

The computer, however, is just as happy solving non-linear equations as linear ones—though it produces a table of numbers or a picture rather than a formula. As computers became more powerful, it was natural to set them loose on non-linear problems. In 1965 Norman Zabusky and M. D. Kruskal were using a computer to study the Fermi–Pasta–Ulam problem, which amounts to solving a particular version of the KdV equations, and they noticed something quite extraordinary. Think of a single-humped travelling wave solution as a single 'pulse'. Now set up conditions in which a large pulse is travelling behind a smaller one, and wait to see what transpires. Well, the larger one moves faster, so it catches up the small pulse and smashes into it. At that point, according to the usual dictum on non-linear equations, everything will go haywire, because you can't just superimpose solutions.

Everything does go haywire. For a time. Then the two pulses miraculously reassemble themselves, with the big one now out in front, and proceed their merry way. This is bizarre. The non-linear waves are behaving, much of the time, like linear ones—except for a period of interaction. But there's a final twist to the tale. Linear waves can also undergo some funny interaction, when it's hard to see which is which. But they separate out in exactly the positions you'd expect them to have reached if there were no interaction at all. The KdV waves don't do this: they undergo a 'phase shift' in their relative positions, as if one of them gets speeded up by the interaction.

Zabusky and Kruskal felt that this remarkable empirical observation demanded theoretical investigation as something worth studying in its own right; and they coined the term *soliton* to mean a solitary wave which retains its individuality even if it collides with other waves. Already in the name we find the vision of a possible connection with fundamental particles, almost all of which also end with 'on'. Electron, muon, pion, gluon. The vision may or may not turn out to have been prophetic.

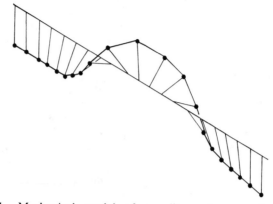

Fig. 34. Mechanical model of a soliton. A row of identical pendulums hanging from a rod, linked by elastic and given a single twist. The soliton is the localized region of twist: it can stay fixed or travel along the rod.

An explanation of the curious properties of the KdV equation came from Kruskal, R. M. Miura, C. S. Gardner, and John Green, who showed that they are in a sense linear equations in disguise—but a new kind of disguise! They form what is now called a 'completely integrable Hamiltonian system'. Other important features, especially the connection with inverse scattering theory, were brought to light by Peter Lax. Many other equations with similar properties and important physical applications were found. A whole mass of physics and mathematics—scattering theory, lattice dynamics, Kac–Moody algebras, Verma modules, cohomology theory, Pontryagin numbers—suddenly entered a new and intimate relationship. There is a good reason for the existence of explicit solutions to KdV-type non-linear equations, and the 'coincidence' *does* lead to a general theory after all. It's easy to say so with hindsight, but coincidences in mathematics *always* deserve explanation. They don't—so to speak—occur merely by coincidence.

The unflattenable bubble

Round about this point, the topologists got in on the act. They wanted to know how solitons could be as stable as they obviously were. The orthodox view had been that solitary waves ought to disperse for reasons of energy. It takes a certain amount of energy to hold a wave together, a certain amount to pull it apart. For a solitary wave, these effects must balance exactly, which seems unlikely. But solitons *don't* disperse, so the argument must be wrong. The mistake can be traced to too restricted a diet of linear mathematics. And a good way to think about non-linear mathematics is to start with the topology.

A physical model for a soliton makes it easier to see what's involved. Imagine an infinite rod, with identical pendulums hanging down at regularly spaced points, like a fringe. Now take infinitely many bits of elastic and link the bob of each pendulum to its immediate neighbours. The role of the elastic is to make each pendulum try to do the same as its neighbours; but any given pendulum can be an individualist provided it can find the energy to stretch the elastic. With this set-up, obviously all pendulums hang down and nothing interesting happens. But suppose we choose a point on the rod, grab hold of the pendulums on one side of that point, and give them all a 360° rotation. Then 'at infinity' they still all hang down; but there's a region of rod where they go through a 360° rotation. If the elastic is not too strong, this region will be fairly short. Its exact shape is determined by the balance between gravity and elastic forces.

The twist is a soliton. You can make it move along the rod just by giving it a tiny shove. The symmetry of the whole system then makes sure it keeps travelling—assuming no friction to slow things down, which is what physicists mean by that word 'Hamiltonian'. We can now see that it's no coincidence that the energy balance is maintained. The elastic forces are trying to hold the wave together; gravity is trying to pull it apart. If you wind the pendulums too loosely, gravity will beat elasticity, and tighten up the twist. But, as it tightens it, the elastic forces increase! So eventually they become equal to the gravitational ones. Similarly, if the twist is too tight, the elastic beats gravity and unwinds it a bit—but then the elastic forces *decrease* until they take the same value as gravity. The whole system has a beautiful feedback mechanism which automatically *adjusts* the two forces until they become equal.

The system *has* to adjust like this because the twist can't be removed altogether. However you move the pendulums around, they have to wrap exactly once round the rod. This is a purely topological property, preserved by continuous changes, and there's no way to avoid it. The total amount of twist is conserved. So the soliton owes its stability to a topological constraint.

Another example of a soliton is the bubble that always appears when you're putting up wallpaper. You can chase it round the wall, but you can't flatten it. In practice it reaches the edge of the paper and escapes, but if you were laying an infinite sheet of wallpaper you'd never be able to get rid of it. This time the two conflicting forces are the pressure in the bubble and the elasticity of the paper.

Particle physics

At this point a quick trip through modern particle physics is in order. In the early days of atomic theory, there were only three types of particle: protons, neutrons, and electrons. Each had a definite mass and an electric charge. The charge was $+1$ for the proton, 0 for the neutron, and -1 for the electron. The total charge on an atom was found by adding the contributions of its constituent particles; and since the total charge was zero, this meant that each atom contained the same number of electrons as it did protons. You couldn't find anything with charge $\frac{1}{2}$, say: it was always a whole number. Electrical charge was a *quantum number* (it came in definite-sized whole number chunks) and it was *conserved* in interactions between particles (the total charge didn't change). The picture soon became more complicated, with the discovery of other kinds of particle, such as neutrinos and mesons. Each particle turned out to have its own antiparticle, which would annihilate it in a burst of pure energy should the two encounter each other. In the *New Yorker* for 10 November 1956 H. P. Furth published a poem in celebration of this effect. Dr Edward Teller is, of course, the father of the American H-bomb, and the A.E.C. is the Atomic Energy Commission.

> Way up beyond the tropostrata
> There is a region dark and stellar

Where, on a streak of anti-matter,
Lived Dr. Edward anti-Teller.

Remote from Fusion's origin,
He lived unguessed and unawares
With all his anti-kith and kin,
And kept macassars on his chairs.

One morning, idling by the sea,
He spied a tin of monstrous girth
That bore three letters: A.E.C.
Out stepped a visitor from Earth.

Then, shouting gladly o'er the sands,
Met two who in their alien ways
Were like as lentils. Their right hands
Clasped, and the rest was gamma rays.

The electron's antiparticle is the positron, the proton's is the antiproton. As physicists bashed particles into each other at higher and higher energies in their accelerators, more and more 'fundamental' particles appeared: kaons, pions, the omega-minus, the J-particle. But there is some kind of method to this madness: a whole host of different quantum numbers, conserved in some interactions but not in others, such as charge, spin, isotopic spin, hypercharge, strangeness, and charm. The most advanced theories today assemble all of these particles from a new kind of basic constituent, called a *quark*; but there are still an awful lot of types of quark. And while each new discovery of a predicted particle is trumpeted world-wide as confirmation of the theory, the six or more discoveries of particles that don't seem to fit the theory at all are seldom mentioned . . . The physicists seem delighted, and they've certainly made some astonishing progress, but I must admit that to me it all looks a bit chewing-gum-and-stringy compared to the elegant beauty of, say, General Relativity. Ptolemy's epicycles fitted experiment pretty well too, but he kind of missed the point. Particle physicists devote a lot of effort to squeezing the next few hundred million dollars out of unsuspecting governments, in order to build the latest hyper-whatsitron, whose ever-higher energies will reveal yet more incomprehensibly weird states of matter. Mathematicians may perhaps be forgiven for wondering whether the interests of science would be better served if there were a ten-year moratorium on building new accelerators, during which time the physicists tried to make proper sense out of what they've already observed. To be fair, some of them are doing just that, as we'll see.

Twists in the fabric

Back to our rod with its dangling pendulums. We can construct other types of soliton, called *kinks*, by twisting the row of pendulums several times round the central rod. There are kinks that wrap round twice, three times, or more—and in either the clockwise or anticlockwise direction. When two kinks collide, these numbers add together: the *total* number of twists does not alter. This is reminiscent of particles, whose electric charge is always a whole number, positive or negative; and the charges add when particles combine. This means that each kink has its own *antikink*, which will annihilate it in a collision, leaving just an untwisted row of undisturbed pendulums. The antikink has the same number of twists, but in the opposite direction. Again this is reminiscent of particles, where for example an electron (charge -1) will annihilate a positron (charge $+1$). So the number of twists in a kink is called its *topological charge*.

Mathematicians have invented a whole range of topological invariants, of which the number of twists is just one. They all share the feature of discreteness: they are measured by whole numbers. They have an algebraic 'additive' structure by which they combine in a natural way. So topology might be able to explain, not just the stability of solitons, but their associated 'quantum numbers' too. All of which suggests a novel picture of the paradoxical features of fundamental particles. Just like solitons, they resemble true particles when isolated from each other, but interact like waves. Just like solitons, they are extremely stable (except during interactions and ignoring short-lived particles that can occur during such interactions). Just like solitons, they possess whole-number invariants that combine according to regular rules. Just like solitons, they possess antiparticles that can annihilate them altogether.

The picture that this paints is one of particles as tiny topological twists, or knots, in the fabric of space-time. They are stable because the topology stops them being untwisted; and their interactions are governed by discrete numbers that reflect properties of their topology. Unfortunately, nobody has been able to make this picture work properly—yet. That's because the KdV equations, on which all of this speculation has been founded, correspond to particles living in a one-dimensional space. But real particles have to live in a three-dimensional space. There's no problem extending the idea of a topological twist into three dimensions; it was done for as many dimensions as you like, long ago. The difficulty is that nobody knows how to write down specific equations, like the KdV equations but consistent with what is already known in physics, that operate in three spatial dimensions and have soliton-like solutions.

Instantons

A certain amount along these lines is known, however. A physicist would call the line of pendulums a 'one-dimensional field theory'. It's one-dimensional because the position of each pendulum can be given by a single number, its distance along the rod. It's a field theory because, associated with each pendulum there is a physical variable, here the angle of rotation, that describes its state. The motion of the entire system is governed by a field equation that holds for each individual pendulum, subject to a coupling between neighbouring pendulums, due to the elastic. There are important one-dimensional field theories in physics, that closely resemble our line of pendulums. The most famous is the *sine-Gordon Equation*, which bears this name because it is obtained from an equation due to Oskar Klein and Walter Gordon by changing one term to a sine function. It is the only known equation whose name is a joke. In 1958 Walter Thirring found a one-dimensional version of quantum mechanics with soliton-like particles and antiparticles. Sydney Coleman proved that this model was basically the same as the sine-Gordon Equation.

But for real particles we need a three-dimensional field theory, whose field variable is associated with the position in ordinary 3-space. This involves a idea from Quantum Field Theory called a *gauge field*. Physicists are especially interested in a special class of gauge fields invented by Chen Ning Yang and Robert Mills (discussed further below). Alexander

Polyakov and Gerhard t'Hooft have shown that solitons can be obtained when a matter field is coupled to a Yang–Mills gauge field. A soliton of this type could in principle be observed experimentally, as a fundamental particle. The snag is that the particle would be a magnetic monopole (never yet seen) with a mass too large to be created in the current generation of particle accelerators. This soliton, like those in the KdV equation, is localized in space but exists for all time. But there is another kind of soliton, which is localized in time too. It is like a particle that suddenly winks into existence and immediately winks out again. It is called an *instanton*, and was discovered by a group of Russian physicists, A. A. Belavin, Polyakov, A. S. Schwartz, and Yu. S. Tyupkin. They interpreted it as a transition between states of a particle (which happens in an instant), rather than a particle itself. The mathematics of instantons is just like the mathematics of sine-Gordon fields.

Yang–Mills gauge fields

It would take us too far afield into physics to describe in detail the intensive interaction between physics and mathematics that is currently going on in gauge field theory, and of which the instanton–soliton connection is just one tiny part. But the topic is too important to ignore altogether. The fundamental idea of a gauge field is *symmetry*. For a long time, physicists have known that symmetries are immensely important in quantum theory. The biggest consumers of group theory today are quantum physicists. The idea is that symmetries of the physical system are reflected in its equations. For example, since any point in Euclidean 3-space looks the same as any other point, then the equations of physics should be invariants under translations, rotations, and reflections—the group of all rigid motions of 3-space.

But in addition there are non-geometric symmetries. If all the charges associated with a system of particles are reversed in sign, from positive to negative and vice versa, then the result is another physically possible system of particles. So here we have *charge-reversal* symmetry. Other symmetries have more bizarre effects, 'rotating' a neutron into a proton through some imaginary space of quantum numbers.

The geometric symmetries of space are *global* symmetries,

which must be applied consistently throughout the entire physical theory. But the symmetry transformations can also be applied in different ways at different points. If the changes due to these transformations can be compensated by some physical field, we have a *local symmetry*. Suppose a number of electrically charged particles are distributed in space, and we compute the electric field. Now move the charges and compute the field again: the result does *not* respect any symmetry. But if we add in a magnetic field as well, then the combined system of electrical and magnetic fields can be transformed in such a way that the symmetry is maintained. So the system of two interconnected fields, described by equations discovered by Maxwell in 1868, has an exact local symmetry.

In quantum mechanics the quantities involved are complex numbers, not real numbers: they have a phase as well as a magnitude. If changing the phase (rotating the complex plane) is a local symmetry, then the system is said do be *gauge-invariant*. The name, which is not particularly evocative, goes back to Weyl in the 1920s. His idea was that physical theories should not depend on the unit of length: the original German can equally be translated as 'calibration-invariant', that is, independent of scale. Weyl's theory didn't work, but it would have if he'd used quantum-mechanical phase instead of scale. Yang and Mills developed a gauge-invariant theory for the isotopic spin symmetry, analogous to Maxwell's electromagnetic theory. It is more complicated mathematically, because the group of local symmetries is not as nice as it is in electromagnetism.

During the 1970s and early 1980s physicists studied Yang–Mills fields. Meanwhile the mathematicians, unaware of this, studied what they called 'connections' on 'fibre bundles'. Eventually it turned out that these were pretty much two names for the same thing, and neither subject has been the same since. There has been spin-off in physics, for example methods for solving the Yang–Mills equations using algebraic geometry and topology; and in mathematics. Donaldson's work on 4-dimensional manifolds is just one of the astonishing results to have emerged; others include deep connections between classical algebraic geometry and partial differential equations. The entire area is undergoing an explosive expansion and is without doubt one of the most exciting developments in twentieth-century mathematics.

Six of our dimensions are missing

Another related topic is the search for a Unified Field Theory. According to current ideas, there are four fundamental forces in Nature: gravity, electromagnetism, weak, and strong. All particles interact via these forces. But the equations describing the interactions are quite different in each case. A unified theory would embrace all four forces as different aspects of some single object. Einstein attempted to find a Unified Field Theory, without success. In the 1960s Sheldon Glashow, Steven Weinberg, and Abdus Salam managed to unify the electromagnetic and weak forces. Current research centres around effects of symmetry, the so-called Kaluza–Klein theories, and objects known as *superstrings*. One version sees the universe not as 4-dimensional space-time, but as a 10-dimensional space-time, six of whose dimensions are curled up into a tight ball and therefore not observed in everyday experience. Another prefers 26 dimensions, of which 22 are curled into a ball. From such bizarre mathematical speculations may the twenty-first century's System of the World be fashioned.

20
More ado about knotting

To me the simple act of tying a knot is an
adventure in unlimited space. A bit of string
affords a dimensional latitude that is unique
among the entities. For an uncomplicated strand
is a palpable object that, for all practical purposes, possesses one dimension only. If we
move a single strand in a plane, interlacing it at
will, actual objects of beauty and utility can
result in what is practically two dimensions; and
if we choose to direct our strand out of this one
plane, another dimension is added which provides opportunity for an excursion that is limited
only by the scope of our own imagery and the
length of the ropemaker's coil.
What can be more wonderful than that?

Clifford W. Ashley

Mathematics has a habit of waking up and biting people.
Subject areas can lie dormant for decades, even centuries,
then suddenly erupt in violent bursts of activity. The infusion
of new ideas from another area, a chance observation, a new
twist to an old trick—almost anything can trigger the eruption.

Topology has been a thriving area of research for most
of this century; but a lot of its progress has been towards
more esoteric questions—multidimensional spaces, intricate
algebraic invariants, big internal generalities. Some of the
simpler parts of topology, its humble origins, have been
neglected. Not deliberately, but because the remaining problems in those areas just seemed too difficult.

Low-dimensional topology has woken up with a vengeance,
and is biting everything in sight. Classical knot theory has been
turned upside down. The reverberations are being felt in molecular biology and in quantum physics. Indeed, it is physics

that has supplied much of the infusion of new ideas. The most curious feature of the whole business is that the discovery that triggered it *could* have been made at any time during the past fifty years. Are too many mathematicians just following fashion? Or does it take a special kind of genius to see something so simple that everybody else has overlooked it? A bit of both, I think. But ideas don't just arise of their own accord: there has to be an entry route, a natural line of thought that leads in the right direction, a bridge across the unknown. This particular bridge was so narrow, its entrance so well concealed by bushes and rocks, that a rare conjunction of circumstances was needed to reveal it.

Within five years, it had become a six-lane highway.

Reef or granny?

When you tie a knot, you use a length of rope. Real lengths of rope have ends. Indeed, without the ends, it would be impossible to tie the knots, for precisely the topological reason that we wish to investigate. The topologist, in contrast, has to get rid of the ends, otherwise the knot can escape by reversing the tying process. The simplest solution is to join the ends together, so that the rope forms an endless loop. Mathematical glue is used, rather than another knot, to avoid infinite regress. More precisely, a topologist defines a *knot* to be a closed loop embedded in ordinary three-dimensional space. A *link* is a collection of several such loops: the individual loops are its *components*. These component loops can be twisted or knotted, and—as the name suggests—may be linked together in any way, including not being linked at all in the usual sense. A knot is just a link with one component, so 'link' is a more general concept than 'knot'.

The picture shows some simple knots and links. Experiment will soon convince you that they are all different. The trefoil can't be deformed into its mirror image; the reef can't be deformed into the granny. All of these experimental observations are challenges to topology: *prove* it. As we shall see, this is far from easy; but the pay-off in terms of useful and intriguing mathematics is huge.

As already discussed in chapter 10, two knots or links are *equivalent*—topologically the same—if one can be deformed

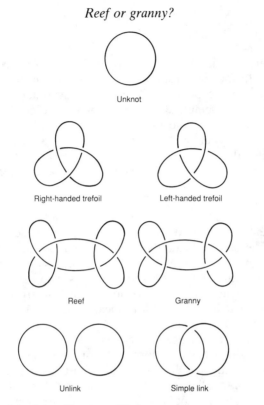

Fig. 35. A variety of knots and links.

into the other by a continuous transformation. From the topological point of view it is unwise to judge knots by appearances: it is possible to deform a knot into an equivalent one that looks very different. Imagine taking an unknotted loop and scrunching it up into a horrible tangle. The tangle looks knotted, but really it isn't. The central problem in knot theory is to find efficient ways to decide whether two given links or knots are topologically equivalent. In particular we want to find out whether something that looks like a knot can be unknotted—that is, is equivalent to the *unknot*, a (manifestly unknotted) circle—and whether a given link can be unlinked.

Knots exist

Are there any knots at all? Perhaps every loop can be un-knotted. In four dimensions, this is true; but in three dimensions, everyday experience strongly suggests otherwise. It is a reasonable mathematical conjecture that genuine knots—not equivalent to the unknot—exist. The first proof was given by Kurt Reidemeister in the 1930s. Why is it so hard to prove something that every boy scout learns at his scoutmaster's knee? Let's see how Reidemeister proved that the trefoil knot can't be untied.

We can represent a knot by its *diagram*: a projection or picture of it, laid flat, with the crossings neatly separated and tiny breaks introduced to show which part crosses over the other. The breaks divide the diagram up into pieces, its *parts*. At any crossing, three parts meet: an overpass and two under-passes. Actually it may sometimes be just two parts, because the overpass may loop round and join the underpass; but I'll continue to talk of three parts, bearing in mind that two of them may be the same.

The same topological knot may have many diagrams, depending on how it is deformed before being projected into the plane. Reidemeister's idea is to keep track of the changes that occur in the diagram when the knot is deformed in space, and try to find a topological invariant: a mathematical pro-perty that is immune to these changes. Any invariant will do, as long as it works. Knots with different invariants are neces-sarily inequivalent. (However, knots with the same invariants may or may not be equivalent, and the only way to decide is either to find an equivalence or to invent a more sensitive invariant.)

Every knot is determined by the crossings in its diagram and the way these join together. If you continuously deform the knot, then you also change its diagram. The pattern of cross-ings usually stays the same, although the shape of the curves joining them will change. But sometimes the crossings them-selves change; and these are the important stages in the defor-mation. Reidemeister proved that the all such changes can be reduced to three different kinds, which I'll call *basic moves*. The first, the *twist*, takes a U-shaped section of string and

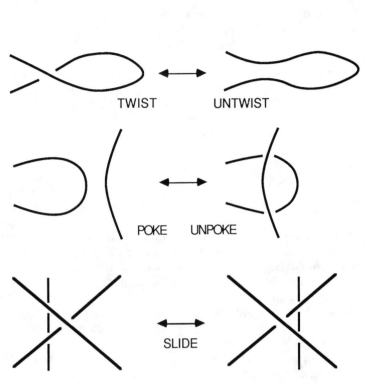

Fig. 36. Basic moves.

twists it over itself to form a loop. It can also be reversed to *untwist* such a loop. The second, the *poke*, pushes a loop of string underneath (or over the top of) another one; and its reverse, the *unpoke*, pulls them apart again. Finally the *slide* just moves one string across the junction of two others. The reverse operation is just another slide in the opposite direction, so we don't need to talk about 'unslides'.

Every continuous deformation of a knot can be represented on its diagram as a sequence of basic moves. Therefore, in order to prove that something is a topological invariant, it is enough to show that each individual basic move leaves it unchanged. In other words, we have to show that it stays the same after a twist or untwist, a poke or unpoke, and a slide.

The property that Reidemester invented is not at all obvious. It is that it should be possible to colour each part of the knot diagram with one of three colours, according to the following rules.

Rule 1   At any crossing, either the colours are all different, or they are all the same (that is, no crossing involves just two distinct colours).

Rule 2   At least two colours are used.

For example, the trefoil knot has a diagram with three parts, and if each is given a different colour, then both rules are obeyed. The unknot has only one part, so rule 2 cannot possibly be obeyed. In short, the trefoil is colourable, but the unknot isn't.

Now comes the cunning bit: the property of 'colourability' is invariant under each type of basic move. If you start with a colourable diagram, and apply a twist, an untwist, a poke, an unpoke, or a slide, then the resulting diagram remains colourable. (It's great fun to work out why. It's no harder than a crossword puzzle, and you don't need to know anything technical: try it.) Put everything together, and you have a proof that you can't unknot a trefoil.

That, in essence, is how Reidemeister solved the problem of the existence of knots. It's a long way removed from wiggling bits of string between your fingers and observing 'experimentally' that they don't come untied. The selection of basic moves, on the other hand, bears a close relation to such experiments: it's a matter of seeing what's important. Proofs are very different creatures from simple-minded experiments, but the germ of the idea must come from somewhere, and experiments often help.

Braid upon the waters

Reidemeister's trick can be generalized. Instead of three colours, you can use four, or five, or some larger number. The rules to be modified, and it's easier to think of integers 0, 1, 2, rather than colours. A knot can be five-coloured, for example, if its parts can be labelled with integers 0, 1, 2, 3, 4, such that

Fig. 37. Colouring a trefoil.

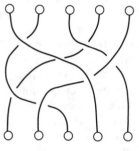

(a)

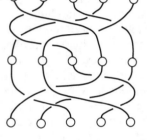

(b)

j   j+1

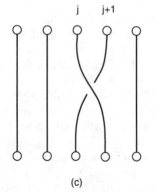

(c)

Fig. 38. (a) A braid. (b) Combining two braids to get a third. (c) Elementary braid with a single crossing.

Rule 1 At any crossing, three parts meet. If the overpass is numbered a and the two underpasses b and c, then we require $2a \equiv b + c$ (mod 5).

Rule 2 At least two numbers are used.

Colourability with a given number of colours is also invariant under basic moves, and hence a characteristic property of a knot. A figure-eight knot can be 5-coloured but not 3-coloured, and a trefoil can be 3-coloured but not 5-coloured, so the figure-eight is topologically different from the trefoil. This kind of game can be played with many different knots, but it has its limitations. It can't distinguish left- and right-handed trefoils; it can't tell reef from granny.

Notice how algebra is creeping into the problem. To begin with, everything was geometric. Now we're talking about numbers, and modular arithmetic. This is typical of topology, and we can here see how such changes of viewpoint occur. The geometry of knots is broken down into parts of a diagram and the way they link; the deformations of knots are replaced by basic moves; and the important feature then becomes the *combinatorics* of parts under the influence of basic moves. This problem in turn is reduced to algebraic questions about labels attached to the parts. This is the topologist's traditional strategy, played with simple ingredients (Reidemeister's colours) or sophisticated ones (the cohomology representation of the mapping class group).

Algebra entered in a big way in 1925 with Emil Artin's invention of the braid group. A *braid* is a system of curves, which start from a line of equally spaced points and end at a parallel line, but winding round each other on the way. Braids are equivalent if you can deform one continuously into the other, just like knots and links; but now the curves have ends, and the ends have to stay fixed; moreover, you're not allowed to push curves over the ends and undo them. They have to stay between the two parallel lines. The new feature is that two braids can be combined, by joining the end of the first to the start of the second. It turns out that braids form a group under this operation. The identity braid consists of parallel curves, not twisted in any way; and the inverse to a given braid is the same braid upside down. Notice that you *only* get a group if deformations of a given braid count as the same braid: the way to cancel out a braid is to combine it with its inverse *and then straighten out the curves.*

Artin found a complete symbolic description of the braid group. Suppose, for example, that the braids have four strands. Every braid can be built up from *elementary braids* s_1, s_2, and s_3 which just swap adjacent points (s_j swapping strands j and $j + 1$). Their inverses are the same braids turned upside down: they look just the same except that overpasses become underpasses. We can symbolically express any braid as a sequence of powers of the s's, using negative powers for inverses: say $s_1^3 s_2^{-4} s_1^7 s_3^5 s_2^{-8}$ and so on. Braids, like knots, may be topologically the same even though they look different, and Artin captured this in the *defining relations* of his braid group, satisfied by 'adjacent' elementary braids:

$$s_1 s_2 s_1 = s_2 s_1 s_2,$$

and so on. He proved that these relations correspond precisely to topological equivalence of braids. That is, suppose two braids are represented as symbol sequences. Then they are topologically equivalent if and only if you can pass from one symbol sequence to the other by applying the defining relations over and over again.

Braids were around before Artin turned them into a group.

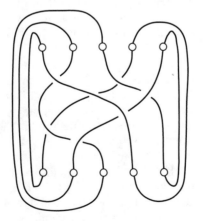

Fig. 39.   Converting a braid to a link.

In particular, in 1923 J. Alexander proved a fundamental connection with links. If you take a braid and join the top to the bottom, you get a link. Alexander proved that every link arises in this way. In short, you can use braids to represent links, and instead of asking questions about links, you can ask them about braids. However, the connection is not entirely straightforward, for many different braids give topologically equivalent links. In 1935 A. A. Markov extended Artin's ideas by finding an algebraic criterion for two braid symbols to yield the same link. There is an awkward extra subtlety: the possibility of changing the number of strands in the braid.

Alexander, as we've already mentioned, invented his own knot invariant, the Alexander polynomial. It's one of the central items in classical knot theory. Like Reidemeister's colouring numbers, it is defined in terms of the knot diagram. In 1936 W. Burau discovered how to find matrices that obey Artin's defining relations. It turns out that the Alexander polynomial of a knot is related to the Burau matrix of the corresponding braid. In short, the Alexander polynomial can be computed algebraically from the braid group. The Alexander polynomial isn't good enough to resolve the great classical questions, such as the inequivalence of left- and right-handed trefoils or the difference between the reef and the granny. However, it leads to more difficult and often highly technical methods that *do* resolve those questions. This 'classical' period of knot theory continued, with the development of more sophisticated invariants, until 1984.

Along came Jones

In that year, a New Zealand mathematician named Vaughan Jones invented an entirely new knot invariant, also a polynomial, but apparently unrelated to anything classical. He published it a year later. It was a watershed in the subject, and we now divide knot theory into two eras: pre-Jones and post-Jones. The Jones polynomial can tell reef from granny as easily as falling off a log, and distinguish left-handed trefoil from right-handed with its hands tied behind its back.

Thing is, Jones wasn't a knot-theorist. He was working in analysis, studying creatures known as von Neumann algebras. If you think of them as groups with bells and whistles on you won't go too far astray, provided you understand that these are very elaborate bells and whistles. Now, just as groups are made from fundamental building-blocks called simple groups, so von Neumann algebras are made from fundamental building blocks called *factors*. Factors come in several kinds: types I, II, and III, with further subdivisions. Jones was thinking about type II_1 factors, which possess a numerical invariant, the *trace*. He focused on a particular set of von Neumann algebras. One of these, for example, has three generators, e_1, e_2, e_3. They satisfy some relations, which include things like

$$e_2 e_1 e_2 = t/(1 + t)^2 e_2,$$

where t is a complex number. D. Hatt and Pierre de la Harpe pointed out that the relations in Jones's algebras are suspiciously similar to those in Artin's braid group; and Jones discovered that if he defined

$$s_j = \sqrt{t}(te_j - (1 - e_j))$$

then the relations become identical. He could then interpret a braid as an element of his algebra, and apply the trace to get a number that varied with t. In fact it's a polynomial in $\sqrt{t}$ and its inverse. The big surprise is that, because of Markov's results, this number is a topological invariant of the link associated with the given braid, a fact that emerged from long discussions between Jones and Joan Birman.

Knowing about the classical polynomial invariants of knots, Jones at first thought that his invariant must be some trivial variation on the Alexander polynomial. But he soon realized that it had to be new. For example, the Jones polynomial for a trefoil is

$$t + t^3 - t^4,$$

whereas that for its mirror image is obtained by replacing t by $1/t$, giving

$$\frac{1}{t} + \frac{1}{t^3} - \frac{1}{t^4}$$

which is manifestly different. The Alexander polynomial is incapable of distinguishing these two knots, and no trivial variation on it can do any better. Similarly the Jones poly-

nomial for the reef is obviously different from the Jones polynomial for the granny. Where classical knot theory had to
push itself to the limit, the Jones polynomial worked effort-
lessly. It was totally new, and extremely powerful. The more
people looked at it, the stranger it became: it wasn't the least
tiny bit like classical knot invariants. What, precisely, *was* it?
And was there anything else out there?

The race was on.

The HOMFLY buzzes

Every so often a scientific discovery is made independently
by several people at the same time. Few cases of multiple
discovery are quite as striking as the one that happened in
October 1984. Within a period of a week or so, no less than
four independent groups in Britain and the United States
obtained the same generalization of the Jones polynomial, and
submitted announcements to the same journal, the *Bulletin
of the American Mathematical Society*. The papers were 'A
new invariant for knots and links' by Peter Freyd and David

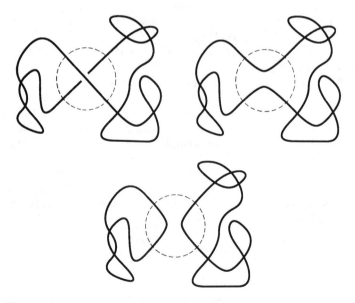

Fig. 40.   Surgery on a knot.

Yetter; 'A polynomial invariant of knots and links' by Jim Hoste; 'Topological invariants of knots and links' by Raymond Lickorish and Kenneth Millett; and 'A polynomial invariant for knots: a combinatorial and an algebraic approach' by A. Ocneanu. As the editors of the journal explained in a half-page footnote, they felt that any attempt to decide priority would be rather pointless; moreover, 'there is enough credit for all to share in'. They persuaded the six authors to combine their results into a single paper, published in April 1985.

In 1970 John Conway had discovered a very quick way to calculate Alexander polynomials, totally different from any classical method. Suppose K, L, and M are three links, that agree except inside a small disk, and differ inside the disk in three possible ways: the strands cross like $\times$, run horizontally like $=$, or vertically like $\|$. Let their Alexander polynomials be $\Delta(K)$, $\Delta(L)$, and $\Delta(M)$, and let t be the variable in the polynomial. Conway proved that

$$\Delta(K) - \Delta(M) + \left(\sqrt{t} - \frac{1}{\sqrt{t}} \right) \Delta(L) = 0.$$

By using this relation repeatedly, it is possible to reduce the calculation of $\Delta(K)$ to combinations of Alexander polynomials of much simpler knots. In this way you can systematically build up a table of Alexander polynomials.

Jones found a very similar formula for his polynomial $V(K)$. The six topologists made the simultaneous discovery by starting from this property of the Jones polynomial, and introducing a new variable into the equation. The resulting polynomial involved two variables rather than one, but nobody minded that. There were technical snags that had to be overcome to obtain a proof that the new variation really worked. Each of the original four papers tackled these technical problems in a different way. The result is often called the *HOMFLY* polynomial (Hoste–Ocneanu–Millett–Freyd–Lickorish–Yetter). It was the first of many: today there are a dozen or more new knot polynomials, each with its own special features; and people are still looking for even bigger and better ones. The great puzzle, in retrospect, is why nobody had discovered any of these polynomials before. This doesn't detract from Jones's achievement: one of the greatest contributions a math-

ematician can make is to spot something so simple and powerful that everybody else has missed it.

Iced Potts

This combinatorial approach to the Jones polynomial makes
it look like a fortunate accident. However, something that
beautiful can't possibly be just a coincidence. Is there a way
to approach the Jones polynomial that makes its topological
invariance inevitable and obvious? Can we find out what it
really is?

Jones, early on, noticed an odd connection with statistical
mechanics, a subject that on the face of it has nothing at all
to do with knots. It arose from physicist's attempts to understand bulk matter—the nature of gases, liquids, and solids.
Matter consists of huge numbers of atoms, rushing around at
random and bouncing off each other. At low temperatures
they stay pretty much in the same place, and combine together
to form a stable solid; as they warm up, they start to move
around more and you get a liquid; when they get really hot,
they fly about all over the place and you get a gas. What is not
obvious from this description is that the *phase transitions*—
changes of state from solid to liquid or from liquid to gas—do
not occur gradually, but suddenly, at very specific temperatures. Imagine warming up a bucket of ice, very slowly, so that
the ice has plenty of time to melt without the temperature
changing much. The ice does not start to melt around $-5\,°C$,
finishing the job by the time the temperature has reached
$+5\,°C$: it changes completely from solid to liquid at a single
temperature, $0\,°C$.

One way to tackle this phenomenon mathematically is using
state models. These are geometric arrangements—usually
lattices—of *sites*. Each site can exist in one of several states;
and it affects the states of its neighbours according to specific
rules. Another important phase transition is magnetism: if you
heat a magnet, it loses all magnetism at a well-defined temperature. State models are more easily understood if we think
of magnetism. Imagine a checkerboard full of tiny magnets,
some pointing up, some down, and each interacting with its
close neighbours. Suppose the interactions are stronger at

lower temperatures, so that neighbouring magnets tend to line up with each other. Then at low temperatures you will get a well-defined direction for the magnetism; but at high temperatures, the magnets point every which way and cancel each other out. There are many such models, and a small number have a remarkable property: you can solve the equations for them exactly and explicitly.

In 1971, long before Jones began his work, H. N. V. Temperley and E. H. Lieb found a connection between two different types of exactly soluble model in statistical mechanics, called Potts models and ice models. In order to establish the connection they had to find a collection of matrices that obeyed particular algebraic rules, determined by the structure of the models. The rules happen to be identical with those of the von Neumann algebras investigated by Jones. In statistical mechanics they are called Temperley–Lieb algebras; and they provide a link between that subject and the new polynomials.

In particular, a knot diagram can be thought of as a state model. Its sites are its crossings, the interactions are determined by the geometry of the knot that joins them. The state of the knot at a crossing is the choice of local structure. This can be of several kinds: overpass on top, overpass underneath (analogous to switching the direction of magnetization at that site), or the overpass can be cut in half and joined to the underpasses to eliminate the crossing altogether. The strange coincidence of the Temperley–Lieb algebras suggests that we may hope to interpret the Jones polynomial in statistical-mechanical terms.

By 1987, following this line of thought, Yasuhiro Akutsu and Miki Wadati did all this, and a lot more: they invented new knot invariants. They remark that 'In most cases where such an interaction between the two fields [of mathematics and physics] occurs, mathematics is usually applied to physics. In this paper, we shall reverse this order.' They then took an idea from physics, the Boltzmann weights of a soluble vertex model, and interpreted it in terms of links, to obtain a new topological invariant.

In the same year Louis Kauffman found a statistical-mechanical interpretation of the original Jones polynomial. It

X   = ||

Fig. 41.   Breaking up a crossing.

begins with yet another new polynomial, which he calls the *bracket polynomial* $\langle K \rangle$ of the knot K. This is defined by three rules:

Rule 1   The bracket of the unknot is 1.

Rule 2   If a separate unknotted component is added to a given link then the bracket multiplies by a variable d.

Rule 3   Select any crossing × in a knot and break it apart in two possible ways = or ||. Then the bracket of the original knot is A times that of the result on replacing × by =, plus B times that of the result on replacing × by ||.

Rule 3 is yet another variation on Conway's formula for the Alexander polynomial. Briefly,

$$\langle \times \rangle = A \langle = \rangle + B \langle \, || \, \rangle.$$

Here d, A, and B are three independent variables.

The choices of 'splitting' at a crossing, ×, =, or ||, can be interpreted as the three 'states' of a statistical-mechanical state model. Kaufmann then derives a formula for his bracket polynomial, as the sum over all states of certain expressions in A, B, and d. In the statistical-mechanical analogy, it is the partition function of the system, a physical quantity from which it is possible to compute thermodynamic variables such as free energy or entropy.

The trick now is to choose A, B, and d to get a topological invariant. By studying the effect of basic moves on the knot diagram, Kaufmann showed that the choice

$$B = A^{-1}, d = -A^2 - A^{-2}$$

makes his bracket polynomial invariant under pokes, unpokes, and slides. However, it is *not* invariant under twists or untwists: it multiplies by $-A^3$ or by $-A^{-3}$. As it happens, there is a second and independent quantity from classical knot theory, the *twist number* or *writhe*, that behaves in exactly the same way. It is a sum over all crossings of terms ± 1, defined according to whether the overpass slants from top left to bottom right or from bottom left to top right. Kaufmann played the writhe off against the bracket to get something that is invariant under all basic moves. If the bracket is multiplied by $(-A)^{-3w}$, w being the writhe, then the result is also invariant under twists and untwists. This modified bracket is thus a topological invariant of knots. It turns out to be a trivial variation on the Jones polynomial. So Kaufmann had found a statistical-mechanical interpretation, in terms of state models, of the Jones polynomial.

The Tait conjecture

The Jones polynomial is a powerful invariant, distinguishing many knots easily where other invariants, such as the Alexander polynomial, fail. A spectacular case is the Kinoshita–Terasaka knot, whose Alexander polynomial is the same as that of the unknot. In contrast, the Jones polynomial distinguishes the two without difficulty. Encouraged by this success, Jones conjectured that his polynomial distinguishes *all* knots that can be constructed from three-stranded braids—if two such knots have the same Jones polynomial, then they are topologically equivalent. For once he was wrong: in 1985 Joan Birman found an infinite number of counter-examples: different three-braid knots with the same Jones polynomial. There are also knots with the same Jones polynomial but distinguished by their Alexander polynomials. Jones is good for some things, Alexander for others.

A long-standing question of classical knot theory is the Tait conjecture. A link is said to be *alternating* if overpasses and underpasses alternate along each component. In 1898 P. G.

Tait conjectured that for an alternating knot, the number of crossings must be the same in any diagram. In 1986 Kunio Murasugi used the Jones polynomial to determine the smallest possible number of crossings in an alternating knot, but at that point he couldn't prove that *every* alternating diagram attains this same minimum number. Before the year was out, Kauffman had used his bracket polynomial to prove the full Tait conjecture. He showed that in any alternating link the number of crossings is one quarter of the difference between the highest power of A that occurs in the bracket polynomial and the lowest power of A that occurs. Since alternating links do not involve the 'twist' basic move, the bracket polynomial does not depend on the choice of diagram; therefore every alternating diagram has the same number of crossings. At about the same time, Murasugi independently proved the same result. A simplified proof was given in 1987 by Turaev.

In 1991 another conjecture by Tait bit the dust: the *flyping conjecture*. A flype is a change made to a knot diagram by flipping over a region (known as a tangle) into which four strands enter. A rather stronger statement than the original conjecture (which applied only to knots) has just been proved

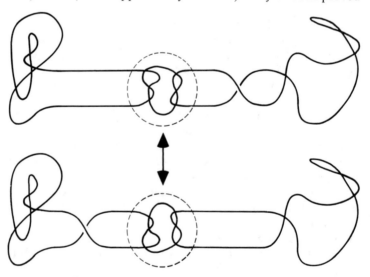

Fig. 42. A flype.

by William Menasco and Morwen Thistlethwaite. It states, with one further technical assumption, that if two alternating links are equivalent, then you can transform one into the other by a series of flypes. This has all sorts of consequences. For example, a knot is equivalent to its mirror image if and only if it can be turned into it by a series of flypes. Since there is no way to flype a trefoil, this gives yet another proof that the left- and right-handed trefoils are different. It also implies that the reef is different from the granny. Indeed, the flyping conjecture implies that most properties of an alternating link must be 'obvious' from its diagram. The proof involves Jones and bracket polynomials, plus some extra ideas from recent three-dimensional topology.

The saga continues. The simplest way to describe what is happening is as a joyous romp through the borderlands of mathematics and physics, with occasional forays into the hinterland in search of a particularly rare specimen. Every time anyone finds something new, it hooks the whole area into another piece of machinery, which can be turned loose on everything currently known. Jones started with an algebra and ended up with an invariant. Within five years, Birman and Hans Wenzl were reversing the process, starting from other people's generalizations of the Jones polynomial and turning them into algebras. Jones pointed out that their new algebra bore a surprising resemblance to something that the algebraist Richard Brauer had worked on in 1937. This fact shed new light on the algebras found by Birman and Wenzl. And so on. It still hasn't stopped fizzing.

Biological links

The new polynomials aren't just intellectual games for mathematicians: they're important in other areas of science, perhaps the most surprising one being molecular biology. Nearly four decades ago James Watson and Francis Crick discovered the 'secret of life', the double-helix structure of the DNA molecule, the backbone on which genetic information is stored and manipulated. The DNA helix is like a two-stranded rope, with each strand twisted repeatedly around the other. When a cell divides, the genetic information is transferred to the new cells by splitting the strands apart, copying them, and joining the new strands to the old in pairs. Anyone who has tried to separate the strands of a long piece of rope knows how tricky this process is: the strands tangle in lumps as you try to pull them apart. In fact DNA is much worse: the helices themselves are 'supercoiled', as if the rope itself has been wound into a coil. Imagine several kilometres of fine thread stuffed into a tennis-ball and you have some idea of how tangled the DNA in a cell must be.

The genetic biochemistry must ravel and unravel this tangled thread, rapidly, repeatedly, and faultlessly; the very chain of life depends upon it. How? Biologists tackle the problem by using enzymes to break the DNA chain into pieces, small enough to investigate in detail. A segment of DNA is a complicated molecular knot, and the same knot can look very different after a few tucks and turns have distorted its appearance.

Until recently biologists had no systematic method for distinguishing these tangled structures, or for working out the associated chemical reactions. The Alexander polynomial, for example, just isn't effective enough. But the new polynomial invariants of knots have opened up new lines of attack, and the topology of knots is becoming an important practical issue in biology. A recent discovery is a mathematical connection between the amount of twisting in the DNA helix and the amount of supercoiling.

Jones's original motivation, in theoretical physics, had no apparent connection with knot theory—and even less with biology. Mathematics is the ultimate in 'technology transfer': a piece of mathematical insight, no matter how it is gained, can be exploited in any area of application that involves the appropriate structures. As the Jones saga shows, original mathematical ideas come from a rich and imaginative environment in which different points of view can interact; they tend not to arise from more specific 'goal-oriented' programmes.

Atiyah's challenge

Most people seem to think that once you have solved a problem, that's it. It's all over. Mathematicians, though, are seldom satified with mere answers; they have an ingrained habit of using the solutions as a springboard for asking new questions. If a calculation miraculously works out with a clean answer, instead of grinding to a halt in the usual morass, most people would breathe a sigh of relief and accept the miracle. Mathematicians, ever suspicious, always want to take the miracle to bits and see what makes it tick. There are good reasons for doing this. Sometimes you find out that it wasn't a miracle at all, but a mistake. Sometimes you find a better way of working miracles. And so it is with the Jones polynomial. You might imagine that mathematicians would be more than satisfied with the amazing series of discoveries that followed on Jones's heels—but of course they aren't. They want more.

There is one curious feature of knot theory in the post-Jones era. A knot, by definition, is an intrinsically three-dimensional animal. But the Jones polynomial and everything that followed it is obsessed with two-dimensional pictures of knots, diagrams in the plane, crossings, overpasses. A truly three-dimensional knot shouldn't 'know about' such things. If you were a knot, you wouldn't recognize that you possessed crossings, any more than we humans go around thinking 'my shadow is a funny shape today'.

At the Hermann Weyl Symposium in 1988, Sir Michael Atiyah laid down two challenges to physicists working in the area of quantum field theory. One was to give a physical interpretation of Donaldson's work on fake four-space. The other was to find an intrinsically three-dimensional approach to the Jones polynomial. Atiyah had chosen well. He knew that if any area of physics would provide answers to these questions, it had to be quantum field theory. There were clues, or bits of clues, all over the place—in the physics, in the mathematics.

One of the important consequences of Donaldson's work is the discovery of new topological invariants for three- and four-dimensional manifolds. Mathematically they are highly unorthodox, but from a physical point of view they can be rationalized after the event as a natural extension of more classical ones. The classical invariants can be interpreted in terms of the solutions to Maxwell's equations for electromagnetism. Solve Maxwell's equations on a given manifold, then play games with those solutions, and out pop the classical invariants (the second homology group and its intersection form, if you want to know). Donaldson's central insight was to replace Maxwell's equations by the Yang–Mills equations, and then play the same sort of game, only more cleverly. In this sense, Donaldson theory was always tied to physics. However, Atiyah was asking whether it is connected with one of the 'hottest' areas of physical research, quantum field theory.

In 1988 Edward Witten found just such a connection, which he called topological quantum field theory. By 1989 he had pushed the same ideas further, and the Jones polynomial was firmly in his sights. From the start, the Jones polynomial was closely related to problems in two-dimensional physics (either two of space or one of space and one of time), such as state models, the Yang–Baxter equation, and Temperley–Lieb algebras. The biggest problem was to impose some kind of order on this unruly area.

A quantum understanding of Donaldson theory involves constructing quantum field theories in which the observables are topological invariants. Witten suggests a loose analogy with general relativity which both explains this point, and reveals the differences involved in the quantum approach. Recall that in general relativity space-time is considered to be curved, and gravitation is a side-effect of this curvature. There are two ways to work with a curved space. One is to use concepts 'intrinsic' to the space, concepts (such as the curvature itself) that do not depend upon a choice of coordinates. The other is to introduce a system of coordinates, and insist that the only meaningful concepts are those that are independent of the coordinate system. Thus, in our own universe it is meaningless to say that Alpha Centauri is two light years east and three light years to the left of the Earth; but fine to say that its distance from Earth is four-and-a-bit light years.

In general relativity, the usual way to proceed is to choose some arbitrary coordinate system, and produce coordinate-free invariants by a kind of 'averaging' over all possible coordinate systems. The coordinate system itself becomes a dynamical variable in the theory. This viewpoint has become ingrained in physics. The mathematicians' approach to the Jones polynomial is identical in spirit. Begin with some two-dimensional projection of the knot ('choice of coordinates') and then show that the answers you get are the same no matter which projection you use (the 'choice of coordinates' makes no difference to the answer). The projection becomes a 'dynamic variable' in the mathematical theory, and your main attention is focused on the significant changes that occur to this variable.

A quantum field theory in which all observables are topological invariants is a quantum field theory that does not depend on any choice of coordinates. By analogy with general relativity and post-Jones knot theory, it would appear that the way to achieve this is to start with some set of coordinates and 'average' over all possible choices. This is technically difficult, because the spaces used in topology are much more flexible than those of general relativity. In general relativity you can bend spaces, so to speak, but not stretch them; in topology, you can do both, and more. Witten's idea was to go against the conventional wisdom, and work intrinsically from the beginning.

I can't go into the details here: you need to be a physicist and a topologist to appreciate them. The main consequence is that there is a formula—once again, a version of the partition function of statistical mechanics, but now in quantum dress—which is defined using a quantum gadget known as a Feynman path integral. This formula is 'obviously' topologically invariant, because everything that you can sensibly write down in Witten's theory *has* to be topologically invariant. The invariance, you recall, is put in right at the beginning. Buried in the partition function of a link is a whole bunch of invariants—the Jones polynomial and its generalizations. The formula works with the link itself, embedded in three-space, independent of any two-dimensional diagram or representation, so it answers Atiyah's challenge.

As a bonus, Witten's approach solves another awkward problem: how to generalize the Jones polynomial to knots that are tied not in ordinary three-space, but in an arbitrary three-dimensional manifold. Just use the same partition function, but define it on the three-manifold rather than on Euclidean three-space. At this point you can even throw the knots away! The result is a set of new topological invariants for three-manifolds, pure and simple.

Witten's theory wraps everything up in a single package: Jones polynomials, braid groups, state models, Yang–Mills equations, Donaldson theory—everything. It offers a totally new, 'intrinsic' way to think about topology in spaces of three and four dimensions. It also provides new insights into quantum field theory, whereby topological ideas can be used to solve physical problems. Within five years of Jones's discovery, the post-Jones era of knot theory has given way to the post-Witten era of low-dimensional topology. Nobody knows where all this will end, but it sure is exciting.

21

Dixit Algorizmi

You say you got a real solution,
Well, you know,
We'd all love to see the plan.
John Lennon and Paul McCartney

Muhammad ibn Musa abu Abdallah al-Khorezmi al-Madjusi al-Qutrubulli was born about AD 810 in Khorezm (now Urgench in the Uzbek SSR). He served as astronomer to the caliph Abd-Allah al-Mamun (The Trustworthy), who inherited the Bayt al-Hikmah (House of Wisdom, or Academy of Sciences) at Baghdad from his father Harun ar-Rashid (The Righteous), and brought it to its peak. Al-Khorezmi's known works are ten in number, and include *Kitab az-zij al-sindhind*, a set of astronomical tables including the first sine and cotangent tables; *Kitab hisab al-'adad al-hindi*, an arithmetic text; and *Kitab al-muhtasar fi hisab al-gabr w'al-muqabalah*, a compact text on the solution of equations. The third book, Latinized as *Ludus algebrae et almucgrabalaeque*, gave us the word 'algebra'. Face it, it could have been worse. The second, Latinized as *Algorithmi de numero Indorum*, brought the Hindu number system to the Arab world, and the resulting Hindu–Arabic number system to Europe. That is, our system of 'hundreds, tens, and units' with a symbol for zero and nine other digits whose value depends on their place. In medieval times, arithmetic was identified with his name, rendered as 'Algorismus'. The formula *dixit Algorizmi* (thus spake Al-Khorezmi) was a hallmark of clarity and authority. His text on arithmetic included all the basic arithmetical processes by which numbers could be added, subtracted, doubled, halved, multiplied, divided, and their square roots extracted. It also had a chapter on business calculations. Today his name survives as the description of any clear and precise procedure for solving a given problem: an *algorithm*. It is a concept at the heart of practical and theoretical computer science, and the mathematics of computability. And you may have noticed that computers, too, have their business uses.

Existence and constructibility

There are many different styles of mathematics. At one, rather nebulous extremity is the pure existence proof, asserting that an object with certain properties must necessarily *exist*, but giving no method to find it. In the middle are more or less constructive techniques which provide a more explicit description of the desired results or objects. At the hard core is the fully constructive algorithm, a perfectly definite procedure guaranteed to calculate exactly what is required, if only you wait long enough.

The problem of transcendental numbers illustrates all three styles well. Cantor's existence proof, mentioned in chapter 7, exhibits not a single transcendental. It merely observes that since there are more reals than algebraic numbers, transcendentals must exist. In the same way, if 349 passengers are booked on to a 348-seater Boeing, they all know that one of them is going to get bumped; but the argument does not prescribe whom. At the intermediate level are transcendence proofs for specific numbers, such as e, π, $2^{\sqrt{2}}$, or Liouville's number. These exploit special properties of the individual numbers involved—just as we can predict that the father travelling with five young children, or the movie star, isn't the one who will be transferred to another flight. An algorithm for transcendence would be a general method of deciding, for any number whatsoever, whether it does or does not satisfy an algebraic equation. No such technique is known; probably no such technique exists. No general 'bumping rule' appears to exist on airliners, either.

There have been extensive philosophical arguments about the value and nature of these types of result. Does a pure existence proof really convey any useful information? One school of modern mathematics, the Constructivists, founded by Errett Bishop in 1967, takes a very restrictive position and refuses to consider any non-constructive arguments at all. My own thinking on this question changed when I was working on applications of computers to algebra. For a particular object (a so-called Cartan subalgebra) there is a classical existence proof, but no algorithm. By approaching the problem from another angle I was able to find a definite algorithm. But . . . my *proof* that the algorithm worked depended on the classical existence theorem! I still have no direct proof. So here an existence theorem is necessary to justify the algorithm. This state of affairs is actually very common. *To calculate something, it is often useful to know in advance that it exists.* Ever since, I've felt that the Constructivists have missed one of the subtleties of our subject.

In fact the waters become murkier still. How should we react to proofs of the *existence of algorithms* that don't actually specify what the algorithm is? I can imagine an algorithm to find existence proofs; or proofs of the existence of algorithms that find existence proofs for algorithms that . . . well, you get the idea.

Algorithms before Algorismus

Algorithmic thinking pre-dates al-Khorezmi by millennia. One of the first significant algorithms is to be found in Euclid's Book Seven, and it illustrates several of their most important features. The problem is to calculate the highest common factor (hcf) of two numbers; that is, the largest number dividing both exactly. Conceptually the simplest solution is to resolve both into primes and retain the smallest power of each prime; and this is the method commonly taught in schools today. For example to find the hcf of 60 and 280 we write $60 = 2^2 \cdot 3 \cdot 5$ and $280 = 2^3 \cdot 5 \cdot 7$ and retain $2^2 \cdot 5$, namely 20, as hcf. But this method is hopelessly inefficient for numbers even of four or five digits, because of the difficulty of factorizing into primes. (Further, the proof that it works *depends* on the Euclidean algorithm!) Euclid's improved method runs as follows. Let the two numbers be m and n, with m the smaller. Then:

(1) Divide n by m with remainder r.
(2) Replace n by m, and m by r.
(3) Repeat from step (1) until the remainder becomes zero.

Then the final divisor m is the required hcf. For instance, to find the hcf of 50983 and 34017 we proceed thus:

50983/34017	leaves remainder 16966
34017/16966	leaves remainder 85
16966/85	leaves remainder 51
85/51	leaves remainder 34
51/34	leaves remainder 17
34/17	goes exactly.

So the hcf is 17. The proof that the algorithm always works is quite easy, and is based on the fact that any number dividing m and n must also divide r, and conversely.

The good-natured rabbit-breeder

The numbers that occur in the Euclidean algorithm decrease rapidly. How rapidly? How many steps does the algorithm take? To answer this we seek the worst case, where the decrease is smallest at each step. Then m must divide n exactly, once at each step. Working backwards we find that m and n should be consecutive terms in the sequence

$$1, 2, 3, 5, 8, 13, 21, 34, 55, 89, \ldots$$

where each number is the sum of the previous two. This sequence has a long history: it was invented by Leonardo of Pisa, nicknamed Fibonacci ('son of good nature') in a problem about rabbit-breeding. The nickname, incidentally, was in vented by Guillaume Libri in the nineteenth century. Leonardo's *Liber Abaci* of 1202 introduced Hindu–Arabic arithmetic to Europe, another link with al-Khorezmi. How big is the nth Fibonacci number? There's an exact formula, but a cruder estimate yields the necessary insight. The number of steps in the Euclidean algorithm is roughly five times the number of digits in the larger of the two numbers involved. Even for 100-digit numbers, no more than 500 steps will be required. We have found a measure of the efficiency of the algorithm, a topic hinted at in chapter 2. Here the 'speed' is proporitional to the number of digits. This is very efficient indeed. In comparison, the length of the 'prime factors' method, using trial division, grows exponentially with the number of digits. For a 100-digit number about 10^{20} steps would be needed.

The Malthusian connection

Suppose we decide on a formal model of computation in which we can define precisely the size of input data and the running time. (One such method is described in the next chapter, but it would be distracting to introduce it here. Informally, think of the input size as the total number of digits involved, and the running time as the number of computational steps.) Then we can measure the efficiency of an algorithm by the way the running time varies with the size of input data. In 1965 J. Edmonds and A. Cobham proposed that the two crucial cases, roughly corresponding to what experience would describe as 'good' and 'bad' algorithms, are polynomial and exponential time. If the running time for input size s grows like a fixed power, say s^3 or s^8, then the algorithm runs in *polynomial time*. If it grows like 2^s or faster then it runs in *exponential time*. Rates of growth sandwiched between these can occur, but are rare in practice. One advantage of this classification is that is does not depend on the chosen model of the computational process. Notice that the 'good' Euclidean algorithm runs in *linear* time (the first power of s), whereas the natural but 'bad' factorization method is exponential.

In 1798 Thomas Malthus wrote a celebrated essay on population pressure, in which he distinguished between the linear growth of food supplies and the exponential growth of the population. It's basically the same distinction, and the crucial point is that in the long run exponential growth will win, however slowly it starts. These crude but effective measures are for theoretical purposes, and must be tempered in practice with other considerations. An algorithm with running-time $10^{500}s^{120}$ is by our definition 'good', but in practice it's useless. One with running-time $\exp(10^{-500}s)$ is 'bad', but might work reasonably for a wide range of values s. But I know of no sensible examples of either of these weird concoctions.

$P = NP$?

Next, let's look at a few typical problems that it would be interesting to find algorithms for.

(1) [Sorting Problem] Given a set of integers, arrange them in ascending order.

(2) [Routing Problem] Work out the best way to route garbage trucks through a city, minimizing mileage subject to constraints such as collecting all the garbage in a working week using no more trucks than the city authorities possess.

(3) [Assignment Problem] Given data on courses, teachers, and students, design a timetable with no clashes.

(4) [Bin-Packing Problem] Given a set of objects of various sizes, fit them into the smallest number of fixed-size bins.

(5) [Travelling Salesman Problem] Given a list of cities, each to be visited exactly once, find the shortest route.

These are all *combinatorial optimization problems*. For each situation X there is a value $c(X)$ that must be minimized. In (1) it is the total number of integers out of order; in (2) the total mileage; and so on. In this context the polynomial/exponential distinction takes on a new and important twist.

Let **P** denote the class of problems that can be solved by an algorithm running in polynomial time: the *easy* problems. A (presumably) more general class, containing most of the interesting problems, is called **NP**, the problems soluble in *non-deterministic polynomial time*. This is more complicated to describe. Suppose for each situation X we are trying to minimize $c(X)$, as above. Suppose for a given number B it is possible to check in polynomial time whether a *solution* X to the problem has $c(X) < B$. Then the problem is **NP**. Note that the presumably hard part—*finding* the solution—is not involved.

For example, consider a jigsaw puzzle. Solving a puzzle with a large number of pieces is hard. But verifying that an attempted solution *is* a solution is very easy. The human eye can do it at a glance; a computer could do it just by checking each piece against its neighbours. Jigsaw puzzles are **NP**.

Clearly every problem in **P** is also in **NP**. Is the reverse true? In other words, if you can *check* a solution in polynomial time, can you *find* it in polynomial time? It sounds unlikely! As we've just seen, you can check the solution of a jigsaw puzzle very rapidly—but nobody expects that to imply that you can also *solve* it rapidly! So we expect to have **P** $\neq$ **NP**. It looks the sort of thing that it should be easy to decide. It's not. It's an unsolved problem, a major challenge that may well have to be bequeathed to the mathematicians of the twenty-first century.

NP-completeness

What makes the **P** ≠ **NP** problem so hard is that it is very difficult to prove that a problem *cannot* be solved in polynomial time. You must envisage *all possible algorithms* for it, and show that each is inefficient. Non-existence proofs are often hard, witness squaring the circle, solving the quintic, proving the parallel axiom, or establishing the continuum hypothesis. Another curious feature is that all of the problems that one might hope are in **NP** but not in **P** are very much on a par with each other, making it difficult to know where to start. Specifically, say that a problem is *NP-complete* if it belongs to **NP**, and if it can be solved in polynomial time, then *every* problem in **NP** can be solved in polynomial time. In other words, if it's not a counterexample, nothing is. Virtually *every* problem anyone can think of in **NP** that isn't obviously already in **P** turns out to be **NP**-complete. Thousands of problems have been studied. And it is also known that the most plausible methods for showing either that **P** = **NP** (simulation) or that **P** ≠ **NP** (Cantor's diagonal argument) cannot be made to work.

Is it true? Is it false? Is it neither, like the continuum hypothesis? Is its truth independent of formal set theory? Nobody has the foggiest idea. It's one of the biggest unsolved mysteries in mathematics.

Bins and boots

Suppose you are given a number of objects, of various sizes, and have to pack them into identical bins, each capable of holding only a certain total size. How many bins do you need? There are a few obvious remarks.

Each object must be smaller than the bins, otherwise there is no solution. You can't get a quart into a pint pot. Similarly the total size of objects, divided by the bin capacity, places a lower bound on the number of bins needed. Aside from these simple observations (which don't help very much) it's hard to think of anything except trying all possible packings and selecting the best. This is, of course, slow and inefficient, and indeed the problem is **NP**-complete. In the absence of optimal solutions, one can ask for less: is there a practical way to find a 'reasonably good' packing? In 1973 Jeffrey Ullman showed that the 'first fit' algorithm (number the bins, and fill as much of each bin as you can before using one with a higher number on it) gives no worse than 17/10 of the optimal number of bins. Moreover, 17/10 is as good as you can get, in general. If the objects are dealt with in order of decreasing size, the 'first fit decreasing' algorithm, the ratio can be improved to 11/9, so you'll never be more than 20 per cent wrong. Anyone packing the boot (USA: trunk) of the family car for a holiday knows that it's best to fit the largest suitcases in first.

A simpler simplex and a torn knapsack

The theory of algorithms is a very new branch of mathematics, and it is changing at high speed as more emphasis is placed on the practical cost of computational methods. One important subject that makes everyday use of computations is *linear programming*, which solves problems like this: how to allocate resources between the production of different items, subject to various price constraints, in order to maximize profits. Obviously such a method has great commercial value. The traditional algorithm was invented by George Dantzig in 1947 and is called the simplex method. It usually works pretty quickly in practice, but for some pathological examples it runs very slowly. The topologist Stephen Smale proved a theorem a few years ago that made this statement precise. In 1979 a lot of publicity was given to a new algorithm devised by L. G. Khachian—possibly because the Press confused the linear programming problem with the much harder travelling salesman problem. Khachian solved the problem of linear programming in polynomial time (the fifth power of the input size, to be precise) by exploiting geometric properties of ellipsoids. As if to emphasize that the distinction between exponential and polynomial time is a little academic, it turned out that in practice Khachian's method is slower than the simplex method—almost all the time! However, in 1985 Narendra Karmarkar managed to combine the best features of both, into a truly efficient polynomial-time algorithm. This is of serious practical importance on the dollars-and-cents level.

Another recent breakthrough involves the knapsack problem. Suppose you have a knapsack, of fixed size, and a collection of objects whose sizes are given. Does any subset of the objects fill the knapsack exactly? In 1982 Ralph Merkle set up a public-key cryptosystem, of the type discussed in chapter 2, based around the knapsack problem; and he bet $1,000 that the resulting code was uncrackable. Adi Shamir promptly broke one version of the code, but that wasn't enough to win the bet, although—to Merkle's irritation—the Press seemed to imagine it was. In 1985 Ernest Brickell broke the whole thing, by developing a fast algorithm to solve the knapsack problem.

22
The limits of computability

Achilles: Hmm . . . there's something just a little peculiar here. Oh, I see what it is! The sentence is talking about itself! Do you see that?

Tortoise: What do you mean? Sentences can't talk.

Achilles: No, but they REFER to things—and this one refers directly—unambiguously—unmistakably—to the very sentence which it is!

Douglas Hofstadter

For centuries mathematicians beat their brains out trying to prove Euclid's parallel axiom. The discovery of non-Euclidean geometries put the problem in a different light. There are geometries different from Euclid's that are equally consistent: if Euclid's geometry is consistent, so are the others. Which raised a much deeper question: *is* Euclid's geometry consistent? Having started from the belief that Euclidean geometry is the true geometry of Nature and that no other geometries can make sense, and being proved wrong on both counts, mathematicians now began to doubt even Euclidean geometry. Is *that* consistent? It was reasonably easy to see that this is so, provided that the real number system is consistent, and to work on the foundations of mathematics after Cantor managed to reduce the consistency problem for the reals first to that for the integers, and then to mathematical logic and formal set theory. Within set theory, and using logic, you can model the integers; with integers you can model the reals; with the reals you can model Euclidean geometry; with Euclidean geometry you can model non-Euclidean geometry. All very well—but what about set theory and logic themselves? It was all a dreadful anticlimax: mathematicians set out to conquer the Universe and ended up doubting the very ground they stood on.

At the turn of the century David Hilbert devised a research programme whose end result was to be a rigorous proof of the consistency of logic and set theory, or equivalently of arithmetic. But before it really got going, and unknown called Kurt Gödel threw a spanner in the works and the programme collapsed in ruins. Gödel showed that there are true statements in arithmetic that can never be proved, and that if anyone finds a proof that arithmetic is consistent, then it isn't! At much the same time, Alan Turing was working in mathematical logic, with a view to clarifying the notion of computability. He too discovered that certain very natural questions *have no answer whatsoever*. By this I do not mean that nobody is clever enough to find the answer: I mean that it can be shown in all rigour that no answer exists at all!

Turing machines

To make our ideas precise—and in this area that's mandatory—we need a precise concept of 'computability'. We might, for example, take an IBM Personal Computer (with an ideal unlimited memory, that is, as many floppy discs as you like), running in assembly language. That would be a bit messy for theoretical purposes, because we'd have to understand the entire operating system of the computer, how its circuits work, and so on. The usual choice is an invention of Turing, who together with John von Neumann provided the theoretical impetus for the creation of the modern computer: it is called a *Turing machine*. It's an idealized computer with the most rudimentary structure possible. Imagine an infinite *tape* divided into square cells, passing under a *head* which can be in a finite number of internal *states*. The head can read what's on the tape, or write symbols 0 and 1 to it. Only 0 and 1 are required, since any information whatsoever can be encoded using just 0s and 1s. For example, use Morse code with 0 representing a dot and 1 a dash. (A standard binary code would be more likely in practice.) The machine's behaviour is controlled by a kind of computer program, called a *Turing-Post program*. This consists of numbered steps, each step being one of seven commands:

Print 1 in the current cell.
Print 0 in the current cell.
Move one cell right.
Move one cell left.
Go to program step k if the current cell holds 1.
Go to program step k if the current cell holds 0.
Stop.

As well as the program, there should be some input data on the tape, for the program to operate on. The machine processes the data according to the program, and then stops.

For example, suppose the initial data is a block of 1s, terminated by 0s, such as

$$\ldots 0111111110 \ldots$$
$$\uparrow$$

where the arrow points to the current cell. Assume that we want the machine to change the terminating 0s to 1, and then stop, thus lengthening the block by two. The following program achieves this:

1. Move right
2. If scanning 1, go to step 1
3. Print 1
4. Move left
5. If scanning 1, go to step 4
6. Print 1
7. Stop

The head moves right until it finds the first 0, and replaces it by a 1; then it moves left until it finds a 0, replaces that with a 1, and stops. You might imagine that such a simple device has only a limited repertoire of tricks. Not so, as Turing showed. Anything that a computer can calculate, so can a Turing machine. Slowly, of course, but that won't bother a theorist seeking insight into the nature of computability.

A Turing-Post program can itself be encoded on the tape, using for example the following codes:

000	Print 0
001	Print 1
010	Move right

011	Move left
10100 . . . 01	Go to step k if 0 scanned
11011 . . . 10	Go to step k if 1 scanned
100	Stop

Here the . . . indicates k repetitions. The program above then encodes as

$$01011010001011110111100011000,$$

and you can reconstruct any program uniquely from its code. Turing showed in 1936 how to construct a *universal* Turing machine, call it U, capable of simulating the action of *any* program in *any* Turing machine. It has a fixed program. Its input data are a sequence of 0s and 1s of the form code $(M) \cdot d$ for a machine with program M and input data d. The program for U scans the code, decodes the actions for M, and carries them out on the data d. Here's another way to say it. Any computer, given enough time and memory, can be progammed to simulate any other computer. A Commodore-64 can pretend to be a Cray. *Very* slowly—speed is hardware-dependent. The program will include a complete coded description of the machine being simulated and the way it works. So, for theoretical purposes, we may as well think of all computations as being performed by Turing machines, and *define* an algorithm to be a Turing-Post program.

The halting problem

An important property of an algorithm is that it should eventually *stop* with a definite answer. A calculation that may go on forever isn't much use, because until it's finished you don't know what the answer is. This raises a natural question. Suppose we are given a Turing-Post program. For some finite strings of input data, it will eventually stop. For others it may not. (For example, 'move right looking for 1s' won't stop if there are no 1s to the right of the head.) The *Halting Problem* is to decide which inputs do or do not cause the machine eventually to stop. In the spirit of the enterprise, we would like to solve this by an algorithm. In other words, we want a Turing-Post program which, given a fixed program P (encoded in 0s and 1s) and variable data d, calculates 0 if P eventually stops for input d, and 1 if not.

Turing was musing along these lines in 1936, and he had a brilliant idea. Suppose such a Halting Algorithm exists, and apply it to the universal Turing machine U with data d. In fact, consider a slightly more complicated program:

(a) Check whether d is the code for a Turing-Post program D. If not, go back to the start and repeat.

(b) If d is the code for D, double the string to get $d \cdot d$.

(c) Use the Halting Algorithm for U with input data $d \cdot d$. If it stops, go back to the start of (c) and repeat.

(d) Otherwise, stop.

Call this program H. You'll have noticed how H gets 'hung up' at stage (a) if d isn't a code of a program, or at (c) if input $d \cdot d$ causes U to halt. Now H is a program, so it has a code h. Does H halt for input h? It certainly doesn't get hung up at (a), because h by definition is a code (of H). It gets past (c) if and only if U does *not* halt with input $h \cdot h$. So, to sum up:

H halts with input data h

if and only if

U does *not* halt with input data $h \cdot h$.

But now consider how the universal machine U simulates a program P. It starts with input data $\text{code}(P) \cdot d$, and behaves just like P with input data d. In other words we have a second 'poem':

P halts with input data d

if and only if

U halts with input data $\text{code}(P) \cdot d$.

Finally put $P = H$, so $\text{code}(P) = \text{code}(H) = h$; and put $d = h$. Then the second poem becomes:

H halts with input data h

if and only if

U halts with input data $h \cdot h$.

But this exactly contradicts the first poem!

This has proved the simplest and most basic undecidability theorem. A problem is *undecidable* if there exists no algorithm to solve it. The undecidability of the Halting Problem places definite limits on the applicability of algorithms. The proof is really an application of Cantor's diagonal argument, like many in undecidability theory. It has a very similar logical structure to a card, on one side of which is written:

THE STATEMENT ON THE OTHER SIDE OF THIS CARD IS TRUE

and on the other:

THE STATEMENT ON THE OTHER SIDE OF THIS CARD IS FALSE.

The used axiom salesman

Turing's theorem, although the most basic undecidability theorem, wasn't the first of its kind. In 1930 Gödel sent for publication a paper which left Hilbert's programme to prove the consistency of arithmetic an unsalvageable wreck. It also scuppered another of Hilbert's most cherished beliefs, that *every* problem can in principle be solved. 'We *must* know, we *shall* know', proclaimed Hilbert. But Gödel proved that there are more things than are dreamt of in Hilbert's philosophy. The two devastating theorems were:

(1) If formal set theory is consistent then there exist theorems that can neither be proved nor disproved.

(2) There is no procedure that will prove set theory consistent.

Legend has it that Hilbert blew his top when he heard of Gödel's results. At first it was thought that the cause lay with the particular formalization of set theory used by Gödel. But as people came to understand how the proof worked, it became clear that the same ideas will do the trick for *any* formalization of arithmetic. It is not any particular axiomatization that is to blame: it is arithmetic itself. With Gödel's theorems the hope of proving beyond any doubt the unsullied purity of mathematics vanished forever. Mathematics was forced to face an ugly fact that all other sciences had come to terms with long before: it is impossible to be absolutely certain that what you are doing is correct. With hindsight this was no bad thing, since the goal of total perfection was obviously a chimera: how can any system demonstrate its own infallibility?

Would you believe someone who said 'I am infallible', and justified it by saying 'because I say so'? (If so, I've got a super used car you'd just *love*, very low mileage; owned by an old lady who only drove it to mathematics seminars on Sundays . . .)

Hilbert's tenth problem

In his address of 1900 to the International Congress of Mathematicians, Hilbert posed twenty-three problems which he felt were of the highest importance. The continuum hypothesis and the consistency of arithmetic were among them. Also included was the search for an algorithm to decide the solubility of arbitrary Diophantine equations. That stayed open until 1970, when Yuri Matijasevich (following earlier work of Martin Davis, Hilary Putnam, and Julia Robinson) showed that no such algorithm can exist. James Jones found an explicit example, a system of eighteen equations of maximum degree 5^{60} in thirty-three variables. Maybe when the counterexamples look like that we're asking the wrong question. For example, what if we insist that the degree be small? Carl Ludwig Siegel found an algorithm in 1972 for equations of degree 2. It is known that no algorithm exists for degree 4, so that leaves cubics as an open problem. Matijasevich's work produced some unexpected spin-offs, notably a polynomial formula whose position values are precisely the primes. It is questionable whether it will have much influence on the theory of primes, though, for it is even more complicated than the system above. Essentially what he did was to find a way to simulate a Turing machine using a Diophantine equation. Consequently anything you can do with Turing machines— compute primes, fail to solve the Halting Problem—can be done with Diophantine equations.

Random sequences

Undecidability theories were pushed to the limit by Gregory Chaitin in 1965 as a by-product of a successful attempt to define what is meant by a *random* sequence of 0s and 1s. A random sequence is of course one that has no pattern. Let's take this oft-made remark seriously, and analyse it. First consider a case with a lot of pattern: a million 1s in a row. You could sit down and write it out in full; but you don't need to, because I've given you a short recipe: 'a million 1s in a row'. With a little more work you could use my definition to set up a Turing machine to output this sequence, with a program which could be encoded as a sequence with far fewer than a million digits. So we could say that a sequence has a *pattern* if it can be computed by a *shorter* sequence, interpreting the latter as the code for a Turing machine. 'A sequence with a pattern has a compact description.' A random sequence is then one that *cannot* be given a compact description. The shortest way to define it is to write it out in full.

Ray Solomonoff anticipated Chaitin's ideas in a less formal way in 1960. Instead of a Turing-Post program, he specified the sequence by a phrase in the English language, and counted the characters (letters, spaces, etc.). Then the sequence

$$11111111111111111111$$

can be defined as 'twenty 1s', using 9 characters; whereas

$$11111111110000000000$$

is 'ten 1s plus ten 0s', requiring 18. Unless more compact descriptions exist, the second sequence is about twice as complicated as the first. Chaitin derived a more formal version of this idea from Information Theory, which was invented by Claude Shannon in the 1940s to study message transmission. It leads to the idea that a given string, or sequence, of 0s and 1s has *complexity* or *information content* c if there is a Turing-Post program which computes the string from suitable data, the total length of program code and data being no more than c. It's now easy to see that there are very few strings having much smaller complexity than their length. Suppose for example the complexity of a string is $n - 5$, where n is the length. Then it can be computed by a machine whose code, plus data, had length $n - 5$ or less. There are at most

$$1 + 2 + \ldots + 2^{n-5} = 2^{n-4} - 1$$

such machines, hence at most 2^{n-4} (let's keep it easy) strings of length n and complexity less than or equal to $n - 5$. Since there are 2^n strings of length n, the proportion having complexity $\leq n - 5$ is at most $2^{n-4}/2^n = 1/16$. Only one string in 16 has complexity only 5 less than its length. Similarly only one string in 500 or so will have complexity $\leq n - 10$. Most strings are random or, at least, pretty close to random.

Chaitin used these ideas to obtain a strong generalization of Gödel's theorems. One consequence can be stated roughly thus. There is a definite number K such that it is impossible to prove that any particular string of 0s and 1s has complexity K. Another consequence is utterly bizarre. While almost all long strings of 0s and 1s are random, by Chaitin's definition, it is impossible to write down an arbitrarily long string and *prove* it to be random. To 'write down' an arbitrarily long string you have to state a general rule for its entries, but then this rule is shorter than suitably large sections of the string, so the string can't be random after all! One way to think of all this is to apply the Solomonoff approach, and consider the phrase 'the smallest string not definable in fewer than 62 characters'— which by its definition is definable in a phrase of 61 characters! This 'Richard Paradox' is at the heart of Chaitin's results.

Another of Chaitin's discoveries is the occurrence of randomness in arithmetic, dramatically extending the results on Hilbert's tenth problem. He has constructed a Diophantine equation requiring 17,000 variables, one of which is a parameter n. Define a number r whose binary expansion has a 0 in the nth place if Chatin's equation has finitely many integer solutions, but a 1 if it has infinitely many. Then r is random. That is, the question 'does Chaitin's equation have finitely many solutions?' depends on the parameter n in a completely random fashion. No finite proof, for example, can answer that question for infinitely many distinct values of n.

Deciding the undecidable

Another mathematician who has thought a lot about issues of computability is Roger Penrose, whose recent book *The Emperor's New Mind* attempts to relate them to quantum physics and the nature of the brain. The central thesis of the book is that nature produces harnessable non-computable processes, but at the quantum level only; and Penrose goes on to speculate that human intelligence may be such a process.

Recent work of Newton da Costa and F. A. Doria casts doubt upon Penrose's thesis, and explores the limits of computability in chaotic classical dynamical systems. They show that undecidability extends to many basic questions in dynamical systems theory. These include whether the dynamics *is* chaotic, whether a trajectory starting from a given initial point eventually passes through some specific region of phase space, and whether the equations are integrable—that is, possess an explicit solution defined by formulas built up from classical functions such as polynomials, sines, and exponentials. Indeed virtually any 'interesting' question about dynamical systems is—in general—undecidable.

This does not imply that it cannot be answered for specific systems of interest: on the contrary, an important recent breakthrough has been the proof by Michael Benedicks and Lennart Carleson that the Hénon system possesses chaotic attractors. However, it does demonstrate the absence of any general formal technique. Another recent 'uncomputability' result in dynamics is Christopher Moore's example of a deterministic system with uncomputable trajectories.

Steven Smale has initiated a systematic theory of computation using real numbers rather than integers. The idea is to model the procedures followed by numerical analysts–for example, solving an equation by finding a sequence of successively better approximations. Smale's set-up corresponds to a computer that can handle real-number arithmetic to infinite precision. No existing computer can actually do this, but theoretical analyses of numerical methods generally assume such a facility. Together with Leonora Blum and Mike Shub, Smale has proved that Julia sets—and, assuming one plausible conjecture in complex analysis, the celebrated Mandelbrot set itself—are *uncomputable*. That this may be so, despite the walls of the planet being papered with computer pictures of Julia and Mandelbrot sets, hinges upon the infinite precision required in the theoretical computations.

A totally different approach also indicates the possibility of curious 'computational' behaviour in classical dynamical systems. You can represent the behaviour of an electronic computer as a dynamical system: just write down idealized but realistic equations for its electronic components. Alternatively, write down equations for a 'real' mechanical Turing machine, with paper tape. Because classical mechanics poses no upper bound upon velocities, it is possible to 'accelerate' time in these equations, so that infinite 'real' time passes within a finite period of 'fake' time. Thus you can concoct a system of classical equations that corresponds to an infinitely fast computer—a Turing machine with a tape that accelerates so rapidly that it can complete an infinite number of operations in one second. Such a machine can in principle compute functions which, in the usual sense of mathematical logic, are uncomputable. For example, it can solve the Halting Problem by running a computation for infinite 'real' time and throwing a particular switch if and only if the program stops. In 'fake' time this entire procedure is over within a fixed period: one then merely inspects the switch to see whether it has been thrown. Additions to conventional mathematical logic that decide undecidable questions are quite respectable things: they are called oracles. So we've invented a classical mechanical oracle for the halting problem. The same oracle can prove or disprove Fermat's Last Theorem by checking all possible sums (infinitely many) of two nth powers. More ambitiously, it can prove *all possible theorems* in a finite period of time, by pursuing *all* logically valid chains of deduction from the axioms of set theory. It is clear from this that some 'classical' dynamical systems simulate bizarre computations. The reason is that perfectly respectable equations have a natural interpretation as models of physically impossible devices. However, it's not easy to implement this informal scenario in a logically impeccable manner—for example, you must actually *write down* the equations, or at least indicate how this could be done; and, more seriously, you have to deal with the potentially infinite-dimensional space of positions on the Turing machine's tape.

Da Costa and Doria use a quite different approach. Their basic idea is inspired by a remark of B. Scarpellini that it may be possible 'to build an analog computer that can simulate functions $f(x)$ so that the predicate $[\varphi(f(x))]$ isn't decidable, while the analog machine itself decides it.' That is, he conceived the existence of an analogue oracle. Developing this idea, da Costa and Doria construct a series of functions $\theta_m(x)$, obtained from classical functions and integrals that involve them. These functions are either identically 0 or identically 1, but their value is formally undecidable. Da Costa and Doria then define a family of classical (that is, non-quantum) dynamical systems, one for each m, each being either a free particle or a simple harmonic oscillator, depending upon the value of θ_m. Since the value of θ_m is undecidable, *there is no formal way to compute which is the case*. However, one can imagine a thought experiment in which an analog device simulates the dynamics for any given m. It will then be immediately apparent to an observer whether or not the motion is simple harmonic or free—in one case the particle oscillates along an interval, in the other it whizzes off to infinity—so the analog machine can effectively decide the undecidable.

This shows that even *classical* systems can go beyond the computational capabilities of Turing machines. The thrust of Penrose's thesis is thus directed elsewhere: if it is to be valid, then it must be posed for 'real' classical systems rather than mathematical ones, a concept that is hard to handle rigorously in a world that is 'really' quantum.

23

The ultimate in technology transfer

> One factor that has remained constant through-
> out all the twists and turns of the history of
> physical science is the decisive importance of
> mathematical imagination. In every century in
> which major advances were achieved the growth
> in physical understanding was guided by a com-
> bination of empirical observation with purely
> mathematical intuition. For a physicist math-
> ematics is not just a tool by means of which
> phenomena can be calculated; it is the main
> source of concepts and principles by means of
> which new theories can be created.
>
> Freeman Dyson

Can we can learn any lessons about the direction in which
mathematics seems to be heading as we tend towards the tail-
end of the twentieth century and plough relentlessly into the
twenty-first? Is mathematics changing? Or is it just the same
tawdry old stuff in a shiny new wrapper?

A bit of both, I think. There is an enormous historical
continuity to mathematics: it would be hard to think of
another subject in which work done four thousand years ago
retains the same freshness that it had on the day it was conceived. The shelf-life of good mathematics is virtually infinite.
Once a problem is solved, it's *solved*; and successive generations can continue to learn from that solution.

Despite which, the manner in which mathematicians go
about their work is definitely changing in important respects.
Let me single out two items for discussion. One is the advent
of the computer and the mathematician's new-found ability to
perform 'experiments'. The other is the interaction between
mathematics and the rest of science. In the first edition of this
book, the final chapter ('A Tour of the Minefield') was a

lengthy diatribe on the relation between 'pure' and 'applied' mathematics. Happily, this is a distinction that increasingly makes no sense at all. As one reviewer put it: 'the minefield does not exist.' Fair enough; but it sure looked as if it did when I wrote that chapter. Maybe it was just a parochial view from one side of the Atlantic. Fortunately, the minesweepers have been busy, the explosives have been defused, the burning oil-wells have been quenched, and sweetness and light reign where once departments fought last-ditch battles over every appointment, every lecture course, every examination question . . .

No, wait a minute, the cease-fire hasn't been as effective as *that* . . .

I can see I'm going to get myself into trouble, so I'll abandon this particular line of enquiry immediately and focus solely on the two issues just introduced. Because this chapter is purely personal opinion (as, I freely admit, was its lengthy predecessor), I'll also keep it short.

Experimental mathematics

Until about the 1970s, computers were seen as glorified calculating machines, 'number-crunchers' whose role was to perform huge calculations. The kind of problems they were useful for were very specific questions in engineering. How does this design of bridge stand up to gales? What are the flight characteristics of a particular wing design on a jet aircraft? How does water flow around the Dutch canal system?

These are eminently important questions in a technological society. However, they hold little to interest the more theoretically minded mathematician, who is interested in general ideas and methods rather than in specific answers. Answers are ten a penny; they only tell you what you wanted to know. The true mathematician always wants to know *more*. Unfortunately, even the simplest general questions can easily lead to calculations far beyond the capabilities of the kind of computer that is entirely happy with supersonic air flow or the gross national product of Lithuania.

The advent of a new generation of computers, with huge memory capacity, very quick arithmetic, and above all precise, high-resolution graphics, has changed all that. It has become possible to use the computer as an experimental tool. Mathematicians can run through large numbers of examples of phenomena that interest them, and see what answers the computer comes up with. Then—the crucial step that turns fun and games into mathematics—they can *think* about the patterns and regularities revealed by those experiments, and see whether they can *prove* that such patterns occur in general.

For example, two minutes with a calculator will convince you that the square of any odd number, minus one, is always a multiple of eight. Thus $3^2 - 1 = 8$, $5^2 - 1 = 24 = 3.8$, $7^2 - 1 = 48 = 6.8, \ldots, 1001^2 - 1 = 1002000 = 125250.8$. But while virtually any mathematician would be convinced by such evidence that there was something interesting to be proved, none of them would consider any such list of numbers, however lengthy, actually to *be* a proof. Here the proof is not far away: $(2n + 1)^2 - 1 = 4n(n + 1)$ and either n or $n + 1$ is even. But many experimental 'facts'—such as Fermat's Last Theorem or the Riemann hypothesis—remain unproven. Others—the Kepler problem, the Bieberbach conjecture—have succumbed only after a lengthy struggle.

There are good reasons for the mathematician's scepticism. Experiments, even extensive ones, can often be misleading. For example, if you count the number of primes of the form $4k + 1$ and $4k + 3$ up to a given limit—say a billion or so—then you will find that those of the form $4k + 3$ always come out ahead. This might seem compelling evidence; but it has been proved that eventually the $4k + 1$ primes catch up. In fact, the lead changes infinitely often. An early theorem along these lines showed that the primes of the form $4k + 1$ gain the lead for numbers no larger than $10^{10^{34}}$. Innocent enough, you may think—but there isn't enough room in the universe to write this number out in normal decimal place notation. Certainly it would never be found by direct trial-and-error experiment.

There is nothing intrinsically radical about the experimental approach. Gauss's notebooks reveal that he performed innumerable calculations trying to guess the answers to tricky problems in number theory. Isaac Newton's papers are also littered with experimental calculations, and so presumably are those of most research mathematicians—after all, their ideas have to come from somewhere. It may be true that occasionally an idea comes fully formed, 'out of thin air'; but most of the time it is the result of a lot of pencil-and-paper 'experiment'.

Two things, however, are new. The first is the computer, which makes it possible to perform large numbers of experiments, or very complicated experiments, to an extent that exceeds the the capacity of the unaided human brain. Computer algebra programs make it possible to perform symbolic calculations—to do algebra just like the brain does, in symbols, not with specific numbers—and to do this with far greater accuracy than the brain, and with much more complicated expressions than the brain can handle. There are 'computer-assisted proofs'—rigorous proofs of theorems that require the computer to do certain boring donkey-work. Appel and Haken's proof of the four-colour theorem was one of the first really spectacular examples.

The second new development is that mathematicians are becoming much more explicit about their use of experiments. They are now being published even when rigorous proofs are lacking. This doesn't represent a drop in standards: everybody agrees that experiments aren't proofs. But the emphasis has shifted, and evidence that may motivate new ideas has become interesting in its own right. In Gauss's day it was all right to perform experiments, but only—so he believed—if you kept them to yourself. He never published his experiments: only the polished proofs that he concocted on their basis.

Nowadays, quite a few people are more interested in the scaffolding than in the building; they want to see in which directions the fox ran. Indeed, a new journal of experimental mathematics has been started, to publish precisely this kind of material. The modern taste is to take a look at the scaffolding as well, in the hope of learning how to make buildings of one's own. Anyone interested in the mental processes of mathematicians is driven to despair by misguided attempts to conceal them. If you are going to train research mathematicians, you need to *show* them the scaffolding, otherwise they get totally the wrong idea of what research is about.

Storm in a Mandelbrot set

As I write, a minor controversy is raging about experimental mathematics. A few people—let me call them the anti-experimentalists—dislike the approach intensely. They believe that it is damaging the integrity of mathematics. A sample of the genre is the statement that fractals—one of the major beneficiaries of the trend towards computer experiments—have added nothing new to mathematics. The anti-experimentalists argue that the theoretical side of fractals is just geometric measure theory or complex analysis, while the applications are not mathematics at all. As for the role of experiment, the anti-experimentalists point out that pictures of the Mandelbrot set make it appear disconnected, whereas in fact it has been proved to be all in one piece.

How naive of the experimental mathematicians to make such a silly mistake.

Except, of course, that they didn't. When Mandelbrot first drew his set, on a rather ancient line printer, he did notice a few tiny specks scattered around, and wondered whether they were dirt or Mandelbrot set. Dirt was soon ruled out. The specks certainly did not appear to be connected to the rest of the set. Did anyone announce that the Mandelbrot set is disconnected? Not at all. As anyone who works with computer graphics soon learns, there may be invisible fine structure below the limits of resolution of the pictures. Instead, Mandelbrot conjectured that the set *connected*, and asked for proof or disproof. It became a well-known *problem* in the field.

Other experimental evidence came from a rather different method of representing the set, by multicoloured contours that focus in upon it. Those contours do *not* surround the specks, as they would if the specks really were disconnected from the rest of the set. Instead, they seem to nestle around thin filaments that run between the specks and the main body of the Mandelbrot set. The filaments themselves do not show up in computer pictures whose pixels are much larger than the filaments are wide, which explains why Mandelbrot didn't see them. Nevertheless, this particular piece of experimental evidence was in their favour. Far from being misleading, it pointed directly at the truth. However, nobody interpreted *those* pictures as proof either. The problem remained open, until John Hubbard and Adrien Douady proved, in the conventional sense of the word, that the Mandelbrot set is indeed connected.

Hubbard has explained the role that the computer pictures played in suggesting a proof. After staring at them for hours, he decided that the set must in fact be connected. Despite naive appearances to the contrary, if you think about what the pictures represent, *and their limitations*, you can convince yourself that the Mandelbrot set must be in one piece. The proof did not emerge directly from the computer pictures, it wasn't even computer-assisted—but the computer experiments played a key inspirational role.

The main plank in the anti-experimentalist platform is the danger of making unwarranted assumptions. In this particular case, however, what actually happened was the exact opposite. It seems to me that their position reduces to the assertion that, if you perform computer experiments badly, or leap to naive conclusions by forgetting the things that computer pictures do *not* show, then you can come up with nonsense. I agree entirely, but I think it's irrelevant. It's true of everything, not just experimental mathematics. If you do it badly, or stupidly, then you get in a mess. In this case, the *only* people who seem to have assumed that the computer pictures indicate that the Mandelbrot set is disconnected are the anti-experimentalists.

Indeed it's hard to see quite what the anti-experimentalists think they are trying to achieve. Most mathematicians are perfectly happy with the use of computers as tools for suggesting interesting theorems. They accept, as do most if not all experimental mathematicians, that the experiments aren't proofs. They remain happy to go out looking for the proofs that the experiments suggest might exist. They neither reject the computer, nor do they believe everything that it seems to be telling them. This is an entirely healthy attitude. It offers experimental mathematics a secure position *alongside* conventional approaches to the subject. As in all branches of science, theory and experiment go hand in hand. Why ever not?

Relations with applied science

One common view of mathematics divides it into two camps, labelled 'pure' and 'applied'. As the century draws to a close, this division is looking increasingly artificial, increasingly dated, and increasingly unhelpful. Applied science has motivated some very important new mathematics, ideas that can if necessary stand on their own as 'pure' mathematics of the highest order. Conversely, innumerable areas of allegedly 'pure' mathematics, invented for their own sake, are turning out to be useful all over the place. The interactions are often remarkable, and sometimes totally unexpected. Only someone with an unusually broad grasp of science and blessed with an unusually fertile imagination might have been able to see them coming.

Virtually every chapter of this book contains examples of interactions in both directions. The application of number theory to secret codes—and the stimulation of new discoveries in number theory. The physical inspiration behind calculus—and the mathematical imagination that turned it into complex analysis. The application of group theory to crystallography—and the new mathematical challenges of quasicrystals. Catastrophes. Chaos. Fractals. Knots.

What we see developing here is not new. It is the way mathematics used to be done, before it got so complicated and so huge that it began to focus itself around specialities, before the specialities became so enmeshed in their own internal problems that they lost sight of their roots in the outside world. And, to the credit of the global mathematical community, it is the way mathematics is starting to face up to the challenges of the next century. Very little of the mathematics discovered over the past fifty years is rendered useless by this trend: on the contrary, almost all *good* mathematics—by which I mean mathematics that is non-trivial and that relates creatively to the rest of the subject—is finding new roles in relation to applied science.

Huge amounts of new mathematics are discovered every year, accompanied by a growing number of new applications, many of them unexpected both in the ingredients that are used and in the areas to which they are applied. They include the use of number theory to study the efficiency of telephone systems, of graph theory to see how best to extract ore from mines, of symmetry to understand the inner workings of electric arc furnaces, and of geometry to improve the production of chemicals. These applications involve new mathematical concepts, new ideas, new theories—they're not just the result of 'putting problems on the computer'. It's very hard to put a problem on the computer without first understanding it in some detail, and computer scientists are increasingly resorting to mathematics to help them sort out how best to use their electronic machinery.

What of the future? Here is my guess at the shape of twenty-first century mathematics. It will be more closely connected to applied science, more explicitly reliant on computer experiments, more aware of its own motivation, less interested in contemplating its own logical navel. It will remain vigorous and innovative. This new *style* of mathematics will continue to be the common language of all branches of science. It will grow organically from mathematics as we know it, it will adapt to meet new challenges, but its pivotal role will be unchanged.

Jovian javelins

Why is mathematics such a powerful and central tool for science? Eugene Wigner wrote about the 'unreasonable effectiveness' of mathematics, and wondered how something that at source was just an artificial construct of the human mind could provide insights into the nature of the universe. Part of his answer was that mathematics is one end of a two-way flow of ideas, with the natural world at the other. I think this is true, but there is an additional feature of mathematics that makes it unusually responsive to the needs of science. That is its universality. Mathematics is about ideas; about how certain facts follow inevitably from others; about how certain structures automatically imply the occurrence of particular phenomena. It lets us build mental models of the world, and manipulate those models in ways that would be impossible in a real experiment.

You want to know how far a javelin would travel in Jupiter's gravity? Plug in the equations, set g to $27\,\mathrm{m\,sec^{-1}}$, and work it out. *You don't have to go to Jupiter and throw one.* (With reasonable assumptions, the current Olympic record of 85.9 metres would become about 33.5 metres on Jupiter.)

Mathematics is about ideas *per se*, not just specific realizations of those ideas. If you learn something about differential equations by analysing Jovian javelins, you may be able to apply that insight to the population dynamics of giraffes or the design of better bicycles. Mathematics doesn't care how you use it, or what problem you apply it to. Its very abstraction— which many people admittedly find repelling—gives it a unity and a universality that would be lost if it were made more concrete or more specialized.

Because many people do find abstraction repelling, it's no good mathematicians thinking that their job is finished when the tool is built. A sales job remains. The world will not beat a path to the mathematical door unless it knows about the improved mathematical mousetrap. Contrary to what many mathematicians appear to assume, the world is not full of people scouring technical journals for a new piece of mathematics that they can put to use in their own area. They need to be told what's available, in their own language, and in the context of their own concerns. I'm not suggesting we all become salespeople; but I am saying that the health of Mathematics Plc depends upon the provision of an adequate sales force.

In short: for the effective transfer of abstract mathematical ideas into science it helps to have an interpreter, a go-between, someone with a foot in both camps. In many areas of science you will find the same underlying ideas—the use of matrices to represent linear equations, say—but bearing the names of many different scientists, one or more per discipline. This is the person who discovered how to use that particular piece of mathematics in that area—or, given the general level of historical accuracy in such attributions, some prima donna who used the idea and attracted attention to it. Because of the lack of communication between disciplines, he or she may well have been reinventing the mathematical wheel at the time the work was first done; nevertheless, it *was* done, and it opened the way to a more organized movement of ideas. There is a lot of talk these days about 'technology transfer', mostly in the sense of taking established technologies and using them in other applications. Mathematics, as I have said before, is the ultimate in technology transfer. It transfers general ideas between *different* technologies.

As a result, mathematics pervades our society. Most of us don't notice this, because it mostly stays behind the scenes. If everything that relied upon it carried a big red label saying MADE USING MATHEMATICS, then red labels would take over from CFCs as a major environmental pollutant. For example, take something as innocuous as a jar of jam. The container is made in a computer-controlled factory. Its label is printed by machinery designed according to mathematical principles of the strength of materials and the behaviour of lubricants. Its ingredients are transported in ships that navigate using satellites, whose orbits are analysed mathematically to give the ship's position. The price is determined by the rules of mathematical economics. Now start thinking about how you make the computers, launch the satellites, build the ships, even make the paper for the label . . .

I've forgotten something . . . oh, yes—the fruit. What could be simpler than raspberries? But even the plants you grow in your garden aren't as straightforward as you may think. Raspberries are disease-prone: they have to be destroyed every five or ten years and replaced by plants imported from colder regions, such as Scotland. The breeders are constantly seeking new varieties, and a great deal of proabability theory and combinatorics goes into the statistical tests used to check the progress of breeding. What about genetic engineering, looming on the horizon? Or the weather forecasting that warns the grower of an early frost?

We live in a complicated world, where nothing is as simple at it once was, nothing is as simple as it seems to be. Mathematics knits that world together.

Don't you forget it.

Further reading

1 THE NATURE OF MATHEMATICS

Donald J. Albers and G. L. Alexanderson (eds.), *Mathematical People—profiles and interviews*, Birkhäuser, Boston, 1985.
E. T. Bell, *The Development of Mathematics*, McGraw-Hill, New York, 1945.
E. T. Bell, *Men of Mathematics* (2 vols.), Penguin, Harmondsworth, 1965.
Carl B. Boyer, *A History of Mathematics*, Wiley, New York, 1968.
Richard Courant and Herbert Robbins, *What is Mathematics?*, Oxford University Press, 1961.
Philip J. Davis and Reuben Hersh, *The Mathematical Experience*, Birkhäuser, Boston, 1981.
Jean Dieudonné, *A Panorama of Mathematics*, Academic Press, New York, 1982.
Matthew P. Gaffney and Lynn Arthur Steen (eds.), *Annotated Bibliography of Expository Writing in the Mathematical Sciences*, Mathematical Association of America, 1976.
Michael Guillen, *Bridges to Infinity*, Rider, London, 1983.
Morris Kline, *Mathematical Thought from Ancient to Modern Times*, Oxford University Press, 1972.
Morris Kline (ed.), *Mathematics in the Modern World*, Freeman, San Francisco, 1969.
Edna E. Kramer, *The Nature and Growth of Modern Mathematics*, Princeton University Press, 1970.
James R. Newman (ed.), *The World of Mathematics* (4 vols.), Simon and Schuster, New York, 1956.
Lynn Arthur Steen (ed.), *Mathematics Today*, Springer, New York, 1978.
Ian Stewart, *Concepts of Modern Mathematics*, Penguin, Harmondsworth, 1975.

2 THE PRICE OF PRIMALITY

Charles C. Bennett, François Bessette, Gilles Brassard, Louis Salvail, and John Smolin, 'Experimental quantum cryptography', *Journal of Cryptology*, to appear.
Artur Ekert, 'Quantum cryptography based on Bell's Theorem', *Physical Review Letters* **67** (1991), 661–3.

Martin Gardner, 'Mathematical Games—a new kind of cypher that would take millions of years to break', *Scientific American*, Aug. 1977, 120–4.

Martin E. Hellman, 'The Mathematics of Public-key Cryptography', *Scientific American*, Aug. 1979, 130–9.

Hendrik W. Lenstra Jr., 'Fast Prime Number Tests', *Nieuw Archiv voor Wiskunde* (40) **1** (1983), 133–44.

Hendrik W. Lenstra Jr., 'Primality Testing', in H. W. Lenstra and R. Tijdeman (eds.), *Computational Methods in Number Theory*, Mathematisch Centrum, Amsterdam, 1982.

Carl Pomerance, 'Recent Developments in Primality Testing', *Mathematical Intelligencer* **3**, no. 3 (1981), 97–105.

Carl Pomerance, 'The Search for Prime Numbers', *Scientific American*, Dec. 1982, 122–30.

Hans Riesel, *Prime Numbers and Computer Methods for Factorization*, Birkhäuser, Boston, 1985.

Manfred Schroeder, *Number Theory in Science and Communication*, Springer, New York, 1984.

Gustavus J. Simmons, 'Cryptology: the mathematics of secure communication', *Mathematical Intelligencer* **1**, no. 4 (1979), 233–46.

S. Wiesner, 'Conjugate coding', *Sigact News* **15** (1983), 78–88.

H. C. Williams, 'Factoring on a Computer', *Mathematical Intelligencer* **6**, no. 3 (1984), 29–36.

Don Zagier, 'The First 50 Million Prime Numbers', *Mathematical Intelligencer* **0** (1977), 7–19.

3 MARGINAL INTEREST

Spencer Bloch, 'The Proof of the Mordell Conjecture', *Mathematical Intelligencer* **6**, no. 2 (1984), 41–7.

Harold Davenport, *The Higher Arithmetic*, Cambridge University Press, 1982.

Underwood Dudley, *Elementary Number Theory*, Freeman, San Francisco, 1978.

Harold M. Edwards, *Fermat's Last Theorem—a genetic introduction to algebraic number theory*, Springer, New York, 1977.

Harold M. Edwards, 'Fermat's Last Theorem', *Scientific American*, Oct. 1978, 86–97.

Noam Elkies, 'On $A^4 + B^4 + C^4 = D^4$', *Mathematics of Computation* (1988).

Eve la Chyl, 'The Weil Conjectures', *Manifold* **16** (1975), 18–28.

L. J. Lander and T. R. Parkin, 'Counterexamples to Euler's Conjecture on sums of like powers', *Bulletin of the American*

Mathematical Society **72** (1966), 1079.

Serge Lang, 'Old and new diophantine inequalities', *Bulletin of the American Mathematical Society* **23** (1990), 37–75.

L. J. Mordell, *Diophantine Equations*, Academic Press, New York, 1969.

Paolo Ribenboim, *13 Lectures on Fermat's Last Theorem*, Springer, New York, 1979.

Nelson Stephens, 'The Prize Puzzle in Mathematics', *New Scientist*, 10 May 1984, 16–19.

André Weil, *Number Theory*, Birkhäuser, Boston, 1983.

André Weil, 'Two Lectures on Number Theory, Past and Present', *l'Enseignement Mathématique* **20** (1974), 87–110.

4 THE NEGLECTED BOOK OF EUCLID

Raymond Ayoub, 'Euler and the Zeta-Function', *American Mathematical Monthly* **81** (1974), 1067–86; **82** (1975), 737.

Albert H. Beiler, *Recreations in the Theory of Numbers—The queen of mathematics entertains*, Dover, New York, 1964.

E. T. Bell, *The Development of Mathematics*, McGraw-Hill, New York, 1945.

Sir Thomas L. Heath (ed.), *Euclid: the thirteen books of the Elements* (3 vols.), Dover, New York, 1956.

Morris Kline, *Mathematical Thought from Ancient to Modern Times*, Oxford University Press, 1972.

Ivan M. Niven, *Numbers: Rational and Irrational*, Random House, New York, 1963.

Carl Douglas Olds, *Continued Fractions*, Random House, New York, 1963.

Alf van der Poorten, 'A proof that Euler missed', *Mathematical Intelligencer* **1**, no. 4 (1979), 195–203.

E. C. Zeeman, 'Research, Ancient and Modern', *Bulletin of the Institute of Mathematics and its Applications* **10** (1974), 272–81.

5 PARALLEL THINKING

Frederic H. Chaffee Jr., 'The Discovery of a Gravitational Lens', *Scientific American*, Nov. 1980, 60–8.

Albert Einstein, 'On the Generalized Theory of Gravitation', in Morris Kline (ed.), *Mathematics in the Modern World*, Freeman, San Francisco, 1969.

George Gamow, 'Gravity', in Morris Kline (ed.), op. cit.

Jeremy Gray, *Ideas of Space*, Oxford University Press, 1979.

David Hilbert and Stephan Cohn-Vossen, *Geometry and the*

Imagination, Chelsea, New York, 1952.
Morris Kline, 'Geometry', in Morris Kline (ed.), op. cit.
P. Le Corbeiller, 'The Curvature of Space', in Morris Kline (ed.), op. cit.
George E. Martin, *The Foundations of Geometry and the Non-Euclidean Plane*, Springer, New York, 1975.
Constance Reid, *Hilbert*, Springer, New York, 1970.
W. W. Sawyer, *Prelude to Mathematics*, Penguin, Harmondsworth, 1955.

6 SPHEREFUL SYMMETRY

J. H. Conway and N. J. A. Sloane, *Sphere Packings, Lattices, and Groups*, Springer, New York, 1988.
Wu-Yi Hsiang, *Sphere Packings and Spherical Geometry—Kepler's Conjecture and Beyond*, preprint PAM-528, Center for Pure and Applied Mathematics, University of California, Berkeley, July 1991.
Wu-Yi Hsiang, *On the Density of Sphere Packings in E^3, I*, preprint PAM-530, Center for Pure and Applied Mathematics, University of California, Berkeley, August 1991.
Wu-Yi Hsiang, *On the Density of Sphere Packings in E^3, II—the Proof of Kepler's Conjecture*, preprint PAM-535, Center for Pure and Applied Mathematics, University of California, Berkeley, August 1991.
Johannes Kepler, *On the Six-Cornered Snowflake*, Oxford University Press, 1966.
J. Leech, 'The problem of the thirteen spheres', *Mathematical Gazette* **40** (1956), 22–3.
C. A. Rogers, *Packing and Covering*, Cambridge University Press, Cambridge, 1964.

7 THE MIRACULOUS JAR

Paul J. Cohen and Reuben Hersh, 'Non-Cantorian Set Theory', in Morris Kline (ed.), *Mathematics in the Modern World*, Freeman, San Francisco, 1969.
Michael Guillen, *Bridges to Infinity*, Rider, London, 1984.
Edward Kasner and James Newman, *Mathematics and the Imagination*, Bell, London, 1961.
James Newman, *The World of Mathematics* (4 vols.), Simon and Schuster, New York, 1956.
Eugene P. Northrop, *Riddles in Mathematics*, Penguin, Harmondsworth, 1960.

Dan Pedoe, *The Gentle Art of Mathematics*, Penguin, Harmondsworth, 1963.
Ian Stewart, *Concepts of Modern Mathematics*, Penguin, Harmondsworth, 1975.
Ian Stewart and David Tall, *The Foundations of Mathematics*, Oxford University Press, 1977.
Leo Zippin, *Uses of Infinity*, Random House, New York, 1962.

8 GHOSTS OF DEPARTED QUANTITIES

Philip J. Davis and Reuben Hersh, *The Mathematical Experience*, Birkhäuser, Boston, 1981.
Marc Diener, 'The Canard Unchained', *Mathematical Intelligencer* **6**, no. 3 (1984), 38–49.
David Fowler, *Introducing Real Analysis*, Transworld, London, 1973.
A. E. Hurd (ed.), *Nonstandard Analysis—recent developments*, Springer, New York, 1983.
M. J. Keisler, *Foundations of Infinitesimal Calculus*, Prindle, Weber, and Schmidt, New York, 1976.
Robert Lutz and Michel Goze, *Nonstandard Analysis*, Springer, New York, 1981.
Larry Niver, *Convergent Series*, Futura Publications, London, 1980.
Abraham Robinson, *Non-Standard Analysis*, North Holland, Amsterdam, 1970.
K. D. Stroyan and W. A. U. Luxemburg, *Introduction to the Theory of Infinitesimals*, Academic Press, New York, 1976.

9 THE DUELLIST AND THE MONSTER

Jonathan Alperin, 'Groups and Symmetry', in Lynn Arthur Steen (ed.), *Mathematics Today*, Springer, New York, 1978.
M. Aschbacher, 'The classification of the Finite Simple Groups', *Mathematical Intelligencer* **3**, no. 2 (1981), 59–65.
F. J. Budden, *The Fascination of Groups*, Cambridge University Press, 1972.
Anthony Gardiner, 'Groups of Monsters', *New Scientist*, 5 Apr. 1979, 37.
Martin Gardner, 'Mathematical Games: Extraordinary non-periodic tiling that enriches the theory of tiles', *Scientific American*, Jan. 1977, 110–21.
Martin Gardner, 'Mathematical Games: The capture of the Monster, a mathematical group with a ridiculous number of elements', *Scientific American*, Jan. 1980, 16–22.
Daniel Gorenstein, 'The enormous theorem', *Scientific American*,

Dec. 1985, 92–103.

Daniel Gorenstein, *Finite Simple Groups—an introduction to their classification*, Plenum, New York, 1982.

István Hargittai, *Quasicrystals, Networks, and Molecules of Fivefold Symmetry*, VCH Publishers, New York, 1990.

Arthur L. Loeb, *Color and Symmetry*, Interscience, New York, 1971.

John Maddox, 'Towards Fivefold Symmetry?', *Nature* **313** (24 Jan. 1985), 263–4.

Lionel Milgrom, 'The Rules of Crystallography Fall Apart', *New Scientist*, 24 Jan. 1985, 34.

Tony Rothman, 'The Short Life of Évariste Galois', *Scientific American*, Apr. 1982, 112–20.

Peter W. Stephens and Alan I. Goldman, 'The structure of quasicrystals', *Scientific American*, April 1991, 24–31.

Ian Stewart, 'The Elements of Symmetry', *New Scientist* **82**, 5 Apr. 1979, 34–6.

Ian Stewart, *Galois Theory*, Chapman and Hall, London, 1973.

Hermann Weyl, *Symmetry*, Princeton University Press, 1952.

10 MUCH ADO ABOUT KNOTTING

Donald W. Blackett, *Elementary Topology*, Academic Press, 1982.

Richard Courant and Herbert Robbins, *What is Mathematics?*, Oxford University Press, 1941, 235–71.

D. B. A. Epstein, 'Geometric Structures on Manifolds', *Mathematical Intelligencer* **4**, no. 1 (1982), 5–16.

P. A. Firby and C. F. Gardiner, *Surface Topology*, Ellis Horwood, Chichester, 1982.

H. B. Griffiths, *Surfaces*, Cambridge University Press, 1976.

David Hilbert and Stephan Cohn-Vossen, *Geometry and the Imagination*, Chelsea, New York, 1952.

John Milnor, 'Hyperbolic Geometry—the first 150 years', *Bulletin of the American Mathematical Society* **6** (1982), 9–24.

Lee Neuwirth, 'The Theory of Knots', *Scientific American*, June 1979, 84–96.

Anthony Phillips, 'Turning a Surface Inside Out', *Scientific American*, May 1966, 112–20.

Peter Scott, 'The Geometries of 3-manifolds', *Bulletin of the London Mathematical Society* **15** (1983), 401–87.

Ronald J. Stern, 'Instantons and the Topology of 4-Manifolds', *Mathematical Intelligencer* **5**, no. 3 (1983), 39–44.

William Thurston, 'Three-Dimensional Manifolds, Kleinian Groups, and Hyperbolic Geometry', *Bulletin of the American Mathematical Society* **6** (1982), 357–81.

William P. Thurston and Jeffrey R, Weeks, 'The Mathematics of Three-Dimensional Manifolds', *Scientific American*, July 1984, 108–20.

Andrew Watson, 'Mathematics of a fake world', *New Scientist*, 2 June 1988, 41–5.

11 THE PURPLE WALLFLOWER

Kenneth Appel and Wolfgang Haken, 'The Four-Color Problem', in Lynn Arthur Steen (ed.), *Mathematics Today*, Springer, New York, 1978.

Kenneth Appel and Wolfgang Haken, 'The Four-colour proof suffices', *Mathematical Intelligencer* **8**, no. 1 (1986), 10–20.

Richard Courant and Herbert Robbins, *What is Mathematics?*, Oxford University Press, 1961.

Philip Franklin, 'The Four Color Problem', *Scripta Mathematica* **6** (1939), 149–56, 197–210.

Frank Harary, 'Some Historical and Intuitive Aspects of Graph Theory', *SIAM Review* **2** (1960), 123–31.

Oystein Ore, *Graphs and Their Uses*, Random House, New York, 1963.

G. Ringel, *Map Color Theorem*, Springer, New York, 1974.

Thomas L. Saaty, 'Remarks on the Four Color Problem: the Kempe Catastrophe', *Mathematics Magazine* **40** (1967), 31–6.

Thomas Saaty and Paul Kainen, *The Four Color Problem*, McGraw-Hill, New York, 1977.

Ian Stewart, *Concepts of Modern Mathematics*, Penguin, Harmondsworth, 1975.

12 SQUAREROOTING THE UNSQUAREROOTABLE

E. T. Bell, *The Development of Mathematics*, McGraw-Hill, New York, 1945.

Lipman Bers, 'Complex Analysis', in *The Mathematical Sciences—a collection of essays for COSRIMS*, MIT Press, Boston, 1969.

Girolamo Cardano, *The Great Art or the Rules of Algebra* (ed. T. R. Witmer), MIT Press, Boston, 1968.

Richard Courant and Herbert Robbins, *What is Mathematics?*, Oxford University Press, 1961.

Philips Davis, *The Thread*, Birkhäuser, Boston, 1983.

Morris Kline, *Mathematical Thought from Ancient to Modern Times*, Oxford University Press, 1972.

Gina Bari Kolata, 'Riemann Hypothesis: elusive zeros of the zeta function', *Science* **185** (1974), 429–31.

Ch. Pommerenke, 'The Bieberbach Conjecture', *Mathematical Intelligencer* **7**, no. 2 (1985), 23–5, 32.
Ian Stewart and David Tall, *Complex Analysis*, Cambridge University Press, 1983.

13 SQUARING THE UNSQUARABLE

V. G. Boltianskii, *Hilbert's Third Problem* (trans. by R. A. Silverman) V. H. Winston, Washington, 1978.
L. Dubins, M. Hirsch, and J. Karoush, 'Scissor congruence', *Israel Journal of Mathematics* **1** (1963), 239–47.
Richard J. Gardner and Stan Wagon, 'At long last, the circle has been squared', *Notices of the American Mathematical Society* **36** (1989), 1338–43.
M. Laczkovich, 'Equidecomposability and discrepancy: a solution of Tarski's circle-squaring problem', *Journal für die Reine und Angewandte Mathematik*, to appear.
C.-H. Sah, *Hilbert's Third Problem: Scissors Congruence*, Pitman, San Francisco, 1979.
Stan Wagon, *The Banach–Tarski Paradox*, Cambridge University Press, Cambridge, 1985.

14 STRUMPET FORTUNE

Rudolph Carnap, 'What is Probability?', *Scientific American*, Sept. 1953, 14–17.
P. Diaconis and B. Efron, 'Computer-Intensive Methods in Statistics', *Scientific American*, May 1983, 96–108.
Paul Halmos, 'The Foundations of Probability', in J. C. Abbott (ed.), *The Chauvenet Papers*, Mathematical Association of America, 1978.
R. Hersh and R. J. Griego, 'Brownian Motion and Potential Theory', *Scientific American*, Mar. 1969, 66–74.
Robert Hooke, *How to tell the Liars from the Statisticians*, Dekker, New York, 1983.
Mark Kac, 'Probability', in Morris Kline (ed.), *Mathematics in the Modern World*, Freeman, 1968.
Mark Kac, 'Probability', in *The Mathematical Sciences—a collection of essays for COSRIMS*, MIT Press, Boston, 1969.
Mark Kac, 'Random Walk and the Theory of Brownian Motion', in J. C. Abbott (ed.), op. cit., 253–77.
Morris Kline, *Mathematics in Western Culture*, Penguin, Harmondsworth, 1972.
A. N. Kolmogorov, 'The Theory of Probability', in A. D.

Aleksandrov (ed.), *Mathematics: its Content, Methods, and Meaning*, MIT Press, Boston, 1963, 229–64.

Oystein Ore, 'Pascal and the Invention of Probability Theory', *American Mathematical Monthly* **67** (1960), 409–19.

W. J. Reichmann, *Use and Abuse of Statistics*, Penguin, Harmondsworth, 1964.

Richard von Mises, *Probability, Statistics, and Truth*, Macmillan, London, 1957.

Warren Weaver, *Lady Luck*, Dover, New York, 1963.

15 THE MATHEMATICS OF NATURE

I. B. Cohen, 'Newton's Discovery of Gravity', *Scientific American*, Mar. 1981, 123–33.

S. Drake, 'The role of Music in Galileo's Experiments', *Scientific American*, June 1975, 98–104.

S. Drake and J. MacLachlan, 'Galileo's discovery of the Parabolic Trajectory', *Scientific American*, Mar. 1975, 102–10.

J. Gerver, 'The existence of pseudocollisions in the plane', preprint, Rutgers University 1990, to appear in *Journal of Differential Equations*.

D. L. Hurd and J. J. Kipling (eds.), *The Origins and Growth of Physical Science* (2 vols.), Penguin, Harmondsworth, 1964.

Morris Kline, *Mathematics in Western Culture*, Penguin, Harmondsworth, 1972.

J. Moser, 'Is the Solar System stable?', *Mathematical Intelligencer* **1**, no. 2 (1978), 65–71.

James R, Newman, *The World of Mathematics*, Simon and Schuster, New York, 1956, especially volume 2.

Richard S. Westfall, *Never at Rest—a biography of Isaac Newton*, Cambridge University Press, 1980.

Z. Xia, 'On the existence of non-collision singularities in Newtonian *n* body systems', doctoral dissertation, Northwestern University 1988.

16 OH! CATASTROPHE!

V. I. Arnol'd, *Catastrophe Theory*, Springer, New York, 1984.

Loren Cobb and Rammohan Ragade (eds.), 'Applications of Catastrophe Theory in the Behavioral and Life Sciences', special issue of *Behavioral Science* **23** (1978), 291–419.

P. W. Colgan, W. Nowell, and N. W. Stokes, 'Spacial Aspects of Nest Defence by Pumpkinseed Sunfish (*Lepomis Gibbosus*): Stimulus features and an application of applied catastrophe theory', *Animal Behavior*, 1980.

Robert Gilmore, *Catastrophe Theory for Scientists and Engineers*, Wiley, New York, 1981.

H. Moysés Nussenzveig, 'The Theory of the Rainbow', *Scientific American*, Apr. 1977, 116–27.

Tim Poston and Ian Stewart, *Catastrophe Theory and its Applications*, Pitman, London, 1978.

P. T. Saunders, *An Introduction to Catastrophe Theory*, Cambridge University Press, 1980.

Ian Stewart, 'Applications of Catastrophe Theory to the Physical Sciences', *Physica* **2D** (1981), 245–305.

Ian Stewart, 'The Seven Elementary Catastrophes', *New Scientist* **68** (1975), 447–54.

Ian Stewart and Martin Golubitsky, *Fearful Symmetry—Is God a Geometer?* Basil Blackwell, Oxford, and Penguin Books, Harmondsworth, 1992.

René Thom, *Mathematical Models of Morphogenesis*, Ellis Horwood, Chichester, 1983.

René Thom, *Structural Stability and Morphogenesis*, Benjamin, New York, 1975.

D'Arcy Thompson, *On Growth and Form* (2 vols.), Cambridge University Press, 1942.

J. M. T. Thompson, *Instabilities and Catastrophes in Science and Engineering*, Wiley, New York, 1982.

Alexander Woodcock and Monte Davis, *Catastrophe Theory*, Dutton, New York, 1978.

E. C. Zeeman, 'Catastrophe Theory', *Scientific American*, Apr. 1976, 65–83.

E. C. Zeeman, *Catastrophe Theory—Selected papers 1972–1977*, Addison-Wesley, Reading, Mass., 1977.

17 THE PATTERNS OF CHAOS

Ralph Abraham and Christopher Shaw, *Dynamics—The geometry of behavior*, Aerial Press, Santa Cruz, 1983, Parts 1–3.

Douglas Adams, *The Hitch-Hiker Trilogy (The Hitch-Hiker's Guide to the Galaxy; The Restaurant at the End of the Universe; Life, the Universe, and Everything; So Long and Thanks for all the Fish)*, Pan Books, London, 1979, 1980, 1982, 1985.

Manfred Eigen and Ruthild Winkler, *Laws of the Game*, Allen Lane, London, 1981.

I. R. Epstein, K. Kustin, P. De Kepper, and M. Orbán, 'Oscillating Chemical Reactions', *Scientific American*, Mar. 1983, 96–108.

James Gleick, *Chaos—Making a New Science*, Viking Press, New York 1987.

Nina Hall (ed.), *The New Scientist Guide to Chaos*, Penguin Books, Harmondsworth, 1992.

Douglas Hofstadter, 'Metamagical Themas—Strange attractors, mathematical patterns delicately poised between order and chaos', *Scientific American*, Nov. 1981, 16–29.

Robert M. May, 'Mathematical Aspects of the Dynamics of Animal Populations, in S. A. Levin (ed.), *Studies in Mathematical Biology*, Mathematical Association of America, 1978, 317–66.

Ilya Prigogine, *From Being to Becoming*, Freeman, New York, 1980.

Ian Stewart, *Does God Play Dice? The Mathematics of Chaos*, Basil Blackwell, Oxford 1989; Penguin Books, Harmondsworth 1990.

J. M. T. Thompson, *Instabilities and Catastrophes in Science and Engineering*, Wiley, New York, 1982.

Philip Thompson, 'The Mathematics of Meteorology', in Lynn Arthur Steen (ed.), *Mathematics Today,* Springer, New York, 1978.

Arthur T. Winfree, 'Sudden Cardiac Death—a problem in topology', *Scientific American*, May 1983, 118–31.

18 THE TWO-AND-A-HALFTH DIMENSION

William Bown, 'Mandelbrot set is as complex as it could be', *New Scientist*, 28 Sept. 1991, 22.

Barry A. Cipra, 'A fractal focus on the Mandelbrot set', *Society for Industrial and Applied Mathematics News*, July 1991, 22.

M. Dekking, M. Mendès France, and A. Van der Poorten, 'Folds', *Mathematical Intelligencer* 4, no. 3 (1982), 130–8; 4, no. 4 (1982), 173–81, 190–5.

Kenneth Falconer, *Fractal Geometry*, John Wiley, Chichester and New York, 1990.

Hartmut Jurgens, Heinz-Otto Peitgen, and Dietmar Saupe, 'The language of fractals', *Scientific American*, Aug. 1990, 40–7.

Benoit B. Mandelbrot, *The Fractal Geometry of Nature*, Freeman, San Francisco, 1982.

Benoit Mandelbrot, interviewed by Anthony Barcellos, in D. J. Albers and G. L. Anderson (eds.), *Mathematical People*, Birkhäuser, Boston, 1985, 206–25.

D. R. Morse, J. H. Lawton, M. M. Dodson, and M. H. Williamson, 'Fractal Dimension of Vegetation and the Distribution of Arthropod Body Lengths', *Nature* 314 (25 Apr. 1985), 731–3.

H. O. Peitgen, D. Saupe, and F. v. Haseler, 'Cayley's Problem and Julia Sets', *Mathematical Intelligencer* 6, no. 2 (1984), 11–20.

Heinz-Otto Peitgen and Peter Richter, *The Beauty of Fractals*, Springer, Berlin and New York, 1986.

J. B. Pollack and J. N. Cuzzi, 'Rings in the Solar System', *Scientific*

American, Nov. 1981, 79–93.
Ian Steward, 'Waves, Wriggles, and Squiggles', *Science Now* **24** (1983), 658–61.

19 THE LONELY WAVE

R. K. Bullough, 'Solitons', *Physics Bulletin*, Feb. 1978, 78–82.
P. G. Drazin, *Solitons*, Cambridge University Press, 1983.
Philip Drazin, 'The life of a solitary wave', *New Scientist*, 30 Nov. 1991, 34–7.
Norman Feather, *Vibrations and Waves*, Penguin, Harmondsworth, 1964.
Daniel Z. Freedman and Peter van Nieuwenhuizen, 'The Hidden Dimensions of Spacetime', *Scientific American*, Mar. 1985, 62–9.
Howard Georgi, 'A Unified Theory of Elementary Particles and Forces', *Scientific American*, Apr. 1981, 40–55.
Banesh Hoffmann, *The Strange Story of the Quantum*, Penguin, Harmondsworth, 1963.
Yuri I. Manin, *Mathematics and Physics*, Birkhäuser, Boston, 1981.
Claudio Rebbi, 'Solitons', *Scientific American*, Feb. 1979, 76–91.
Gerard t'Hooft, 'Gauge Theories of the Forces between Elementary Particles', *Scientific American*, June 1980, 90–114.

20 MORE ADO ABOUT KNOTTING

C. W. Ashley, *The Ashley Book of Knots*, Faber and Faber, London, 1947.
P. Freyd, D. Yetter, J. Hoste, W. B. R. Lickorish, K. Millett, and A. Ocneanu, 'A new polynomial invariant of knots and links', *Bulletin of the American Mathematical Society* **12** (1985), 239–46.
V. F. R. Jones, 'A polynomial invariant for knots via von Neumann algebras', *Bulletin of the American Mathematical Society* **12** (1985), 103–11.
Vaughan F. R. Jones, 'Knot theory and statistical mechanics', *Scientific American* Nov. 1990, 52–7.
W. B. R. Lickorish and K. C. Millett, 'The new polynomial invariants of knots and links', *Mathematics Magazine* **61** (Feb. 1988), 3–23.
L. H. Kauffman, 'Statistical mechanics and the Jones polynomial', *Contemporary Mathematics* **78** (1988), American Mathematical Society, Providence RI, 283–97.
Toshitake Kohno (ed.), *New Developments in the Theory of Knots*, World Scientific, Singapore 1990.
William W. Menasco and Morwen B. Thistlethwaite, 'The Tait flyping conjecture', *Bulletin of the American Mathematical Society*

25 (1991), 403–12.

Ian Stewart, *Game, Set and Math*, Basil Blackwell, Oxford, 1989; Penguin Books, Harmondsworth, 1991.

Andrew Watson, 'Twists, tangles, and topology', *New Scientist*, 5 Oct. 1991, 42–6.

E. Witten, 'Quantum field theory and the Jones polynomials', *Communications in Mathematical Physics* **121** (1989), 351–99.

21 DIXIT ALGORIZMI

'Simplified Simplex', *Scientific American*, Jan. 1985, 42–3.

'Torn Knapsack', *Scientific American*, Jan. 1985, 45.

Robert G. Bland, 'The Allocation of Resources by Linear Programming', *Scientific American*, June 1981, 108–19.

Allan Calder, 'Constructive Mathematics', *Scientific American*, Oct. 1979, 134–43.

Harold Jackson, 'Russian Way with the Travelling Salesman', *Guardian*, 29 Oct. 1979.

Donald E. Knuth, 'Algorithms', *Scientific American*, Apr. 1977, 63–80.

Gina Bari Kolata, 'Mathematicians Amazed by Russian's Discovery', *Science 2*, Nov. 1979, 545–6.

H. W. Lenstra Jr., 'Integer Programming and Cryptography', *Mathematical Intelligencer* **6**, no. 3 (1984), 14–19.

R. Pavelle, M. Rothstein, and J. Fitch, 'Computer Algebra', *Scientific American*, Dec. 1981, 102–13.

G. J. Tee, 'Khachian's Efficient Algorithm for Linear Inequalities and Linear Programming', *SIGNUM Newsletter*, Mar. 1980.

Heinz Zemanek, 'Al-Khorezmi', in A. P. Ershov and D. E. Knuth, *Algorithms in Modern Mathematics and Computer Science*, Springer, New York, 1981.

22 THE LIMITS OF COMPUTABILITY

Lenore Blum, Mike Shub, and Steve Smale, 'On a theory of computation and complexity over the real numbers: NP-completeness, recursive functions and universal machines', *Bulletin of the American Mathematical Society* **21** (1989), 1–46.

Gregory J. Chaitin, 'Randomness and Mathematical Proof', *Scientific American*, May 1975, 47–52.

Barry Cipra, 'Two mathematicians prove a strange theorem', *Society for Industrial and Applied Mathematics News*, May 1991, 1, 7, 9.

N. C. A. da Costa and F. A. Doria, *Classical Physics and Penrose's Thesis*, preprint, Institute for Advanced Studies, University of São

Paulo; to appear in *Foundations of Physics Letters*.

N. C. A. da Costa and F. A. Doria, *Undecidability and Incompleteness in Classical Mechanics*, preprint, Institute for Advanced Studies, University of São Paulo, to appear in *International Journal of Theoretical Physics*.

Martin Davis, 'What is a Computation?', in Lynn Arthur Steen (ed.), *Mathematics Today*, Springer, New York, 1978.

Martin Davis and Reuben Hersh, 'Hilbert's Tenth Problem', *Scientific American*, Nov. 1973, 84–91.

Martin Davis, Yuri Matijasevich and Julia Robinson, 'Hilbert's Tenth Problem. Diophantine Equations: Positive aspects of a negative solution', in *Mathematical Developments Arising from Hilbert Problems*, Proceedings of Symposia in Pure Mathematics XXVIII, American Mathematical Society, Providence, 1976, 323–78.

Douglas R. Hofstadter, *Gödel, Escher, Bach—an eternal golden braid*, Penguin, Harmondsworth, 1980.

John E. Hopcroft, 'Turing Machines', *Scientific American*, May 1984, 86–98.

R. Penrose, *The Emperor's New Mind*, Oxford University Press, Oxford, 1990.

Ian Stewart, *Concepts of Modern Mathematics*, Penguin, Harmondsworth, 1975.

Ian Stewart, 'The dynamics of impossible devices', *Nonlinear Science Today* **1**, no. 4 (1991), 8–9.

L. J. Stockmeyer and A. K. Chandra, 'Intrinsically Difficult Problems', *Scientific American*, May 1979, 124–33.

23 THE ULTIMATE IN TECHNOLOGY TRANSFER

M. F. Atiyah, 'The Unity of Mathematics', *Bulletin of the London Mathematical Society* **10** (1978), 69–76.

William Bown, 'New-wave mathematics', *New Scientist*, 3 Aug. 1991, 33–7.

Freeman Dyson, 'Mathematics in the Physical Sciences', in *The Mathematical Sciences—Essays for COSRIMS*, MIT Press, Boston, 1969.

Steven G. Krantz, 'Fractal Geometry', *Mathematical Intelligencer* **11**, no. 4 (1989), 12–16.

Benoit B. Mandelbrot, 'Some "facts" that evaporate upon examination', *Mathematical Intelligencer* **11**, no. 4 (1989), 17–19.

Brockway McMillan, 'Applied Mathematics in Engineering', in Thomas L. Saaty and F. Joachim Weyl (eds.), *The Spirit and the Uses of the Mathematical Sciences*, McGraw-Hill, New York, 1969.

Yuri I. Manin, *Mathematics and Physics*, Birkhäuser, Boston, 1981.

Arturo Sangalli, 'The burden of proof is on the computer', *New Scientist*, 23 Feb. 1991, 38–40.

Manfred Schroeder, *Number Theory in Science and Communication*, Springer, New York, 1984.

Igor Shafarevich, 'On Certain Tendencies in the Development of Mathematics', *Mathematical Intelligencer* **3**, no. 4 (1981), 182–4.

Lynn Arthur Steen, *Mathematics Tomorrow*, Springer, New York, 1981.

Lynn Arthur Steen, 'Mathematics for a new century', *Notices of the American Mathematical Society* **36** (1989), 133–8.

Ian Stewart, 'The science of significant form', *Mathematical Intelligencer* **3**, no. 2 (1981), 50–8.

Ian Stewart, 'The electronic mathematician', *Analog Science Fiction/Science Fact* **107**, Jan. 1987, 73–89.

Hermann Weyl, 'The Mathematical Way of Thinking', in James R. Newman (ed.), *The World of Mathematics*, Simon and Schuster, New York, 1956.

Alfred North Whitehead, 'Mathematics as an Element in the History of Thought', in James R. Newman (ed.). op. cit.

Eugene P. Wigner, 'The Unreasonable Effectiveness of Mathematics in the Natural Sciences', in Thomas L. Saaty and F. Joachim Weyl (eds.), op. cit.

Index